Ready For 6th Grade

Math

A Summer Learning Activity Workbook For Students Entering Grade 6 [Ages 11-12]

Written and edited by Matthew Bowling

Printed in the United States of America

This paper meets the requirements of ANSI/NISO Z39.48-1992 (Permanence of Paper).

Mometrix offers volume discount pricing to institutions. For more information or a price quote, please contact our sales department at sales@mometrix.com or 888-248-1219.

Paperback
ISBN 13: 978-1-5167-3151-0
ISBN 10: 1-5167-3151-4

TABLE OF CONTENTS

Introduction

Congratulations on finishing 5th grade! That is a big accomplishment, and you should be proud of yourself.

As you get ready to move into **6th grade**, it is normal to feel a little nervous. This workbook is designed to help you feel prepared and confident as you take on new challenges. You've got what it takes to succeed.

This workbook is organized into chapters that review important 5th grade skills and introduce new concepts you'll see in 6th grade. Each chapter starts with a short lesson to refresh your memory or even introduce new things before you start the daily activities. You can always go back to the lesson page if you need extra help.

The workbook is made to be completed over 10 weeks, with each day taking no more than 15 minutes. That is just enough time to practice without feeling overwhelmed. As you work through the chapters, the activities will gradually become more challenging, helping you strengthen your thinking skills and prepare for the work you'll do in 6th grade.

Sometimes you may work with an adult, and other times you may work independently. Either way, you're building skills, confidence, and strong learning habits. Get ready, because 6th grade is going to be a great year!

Chapter 1: Whole Numbers and Operations

Whole numbers include the positive integers and zero. 0, 1, 2, 3... and so on, are all whole numbers. It is very important to be able to **add**, **subtract**, **multiply**, and **divide** whole numbers quickly and easily. Addition, subtraction, multiplication, and division with whole numbers should be automatic.

To add multiple-digit numbers together, start in the ones place and use basic addition rules. When the number equals ten or more, the first digit of the number carries over to the next place value. This is called **regrouping**. Look at the example below. When the numbers in the ones column add up to a number larger than 10, the first digit of that number carries over to the tens column. This same process is used for larger numbers as well.

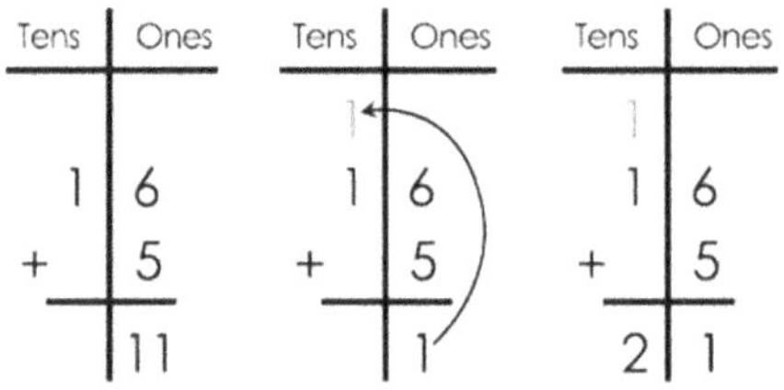

To subtract multiple-digit numbers together, start in the ones place and use basic subtraction rules. If the top number in the column is smaller than the bottom, borrow a group of ten from the column to the left. This is called **borrowing**. Look at the example below. This same process is used for larger numbers as well.

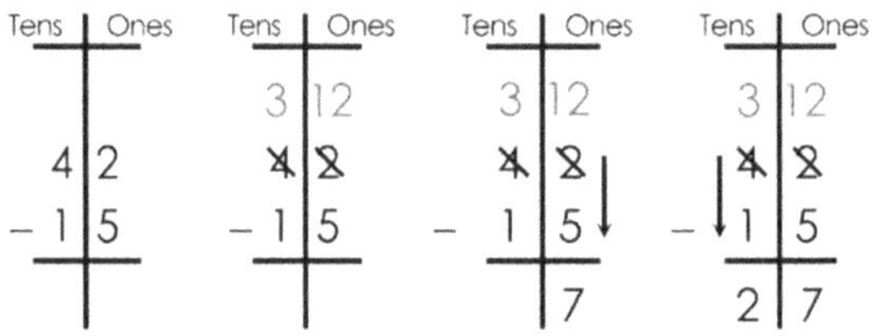

Multiplying large numbers takes practice. When you set up a multiplication problem with large numbers, it is very important to line up place values and understand the concept of **partial products**. Partial products are the pieces of a larger multiplication problem that, when added together, give the whole product. One way to represent $1{,}542 \times 12$ is to break 12 into $10 + 2$ and distribute. We can also use this strategy the other way, where 1,542 is written in expanded form and the 12 is distributed. We can also set up a problem using the standard algorithm.

$$\begin{array}{r} {}^{1} \\ 1{,}542 \\ \times \quad 12 \\ \hline 3\,084 \end{array}$$

$$\begin{array}{r} 1{,}542 \\ \times \quad 12 \\ \hline 3{,}084 \\ 15{,}420 \end{array}$$

$$\begin{array}{r} {}^{1} \quad \\ 3{,}084 \\ +15{,}420 \\ \hline 18{,}504 \end{array}$$

Dividing large numbers takes practice. The most common strategy for large-digit division is using long division. Long division is a shorthand way to divide large numbers by place value. There are a couple of ways to complete long division problems. When using long division, we set up our problems like this:

$$25\overline{)6{,}585}$$

The first step in this problem is to ask how many times 25 goes into 6. 25 is larger than 6, so the answer would be zero. We will put a zero in the thousands place. Now we move to the hundreds place. How many times will 25 go into 65? Twice. We would say twice because $2 \cdot 25 = 50$ and $3 \cdot 25 = 75$; 75 is greater than 65, so we will choose $25 \cdot 2$. Now we multiply $25 \cdot 2$, place this under 65, subtract, and bring down the next digit. Now we ask, "how many times will 25 go into 158?" 6, because $6 \cdot 25 = 150$. So, we place a 6 in the tens place, multiply, subtract, and bring down the last digit. Now, we follow the same process. How many times does 25 go into 85? 3 times. Place a 3 in the ones place, multiply, and subtract. There is nothing else to bring down, but we have 10 left over. This is our remainder, so we can say that $6{,}585 \div 25 = 263$ with a remainder of 10, or $263\frac{10}{25}$. The fraction $\frac{10}{25}$ can be simplified, so in simplest terms, this is $263\frac{2}{5}$.

$$\begin{array}{r} 0\,263 \\ 25\overline{)6{,}585} \\ \underline{5\,0} \\ 158 \\ \underline{150} \\ 85 \\ \underline{75} \\ 10 \end{array} \leftarrow \text{remainder}$$

An **exponent** is used to identify a multiple of the same factor. The symbol for an exponent is a small number to the top right of a whole number. The small number shows how many times the whole number should be multiplied by itself. For example, 2^3 is in exponent form and is equivalent to $2 \times 2 \times 2$ which is expanded form. It is also equivalent to 8 which is standard form. When identifying equivalent expressions using exponents, you may be required to simplify exponents, expand exponents, or generate exponents.

Complete one worksheet each day in this chapter. Each day should take no longer than 15 minutes to complete.

Day 1

8-Digit Addition

Solve the problems below using regrouping.

1.

```
  56,789,103
  10,365,987
+ 37,697,443
____________
```

2.

```
  12,521,356
  69,126,078
+ 68,703,491
____________
```

3.

```
  78,417,245
  15,467,909
+ 81,346,579
____________
```

4.

```
  10,567,892
  72,431,965
+ 69,743,101
____________
```

5.

```
  60,567,986
  91,265,347
+ 13,258,924
____________
```

6.

```
  34,621,937
  76,864,344
+ 52,843,321
____________
```

DAY 2

8-DIGIT SUBTRACTION

Solve the problems below using regrouping.

1.

```
  32,445,311
- 19,306,805
____________
```

2.

```
  54,009,812
- 23,329,605
____________
```

3.

```
  72,511,420
- 49,964,122
____________
```

4.

```
  43,652,019
-  7,994,364
____________
```

5.

```
  17,457,333
-  9,864,127
____________
```

6.

```
  85,674,152
- 39,200,997
____________
```

7.

```
  55,925,321
-    365,544
____________
```

8.

```
  90,899,425
-  1,621,678
____________
```

Day 3

Multiplication Word Problems

Use multiplication to solve the problems below.

1. George has lots of pigs. He keeps 147 pigs in each pen. He has 23 pens. How many pigs does he have in all?

2. Penny is cooking muffins. She can cook 4 muffins per pan. She has 239 pans. How many muffins can she cook?

3. Mickey is going trick or treating. If he goes to 128 houses and collects 12 pieces of candy at each house, how much candy will Mickey have?

4. Nathan is a talented basketball player. If he scores 19 points in each of his next 32 games, how many points will he score?

DAY 4

MULTIPLYING 4-DIGIT NUMBERS BY 2-DIGIT NUMBERS

Solve the problems below.

1. $\begin{array}{r} 2,631 \\ \times \quad 12 \\ \hline \\ + \quad \\ \hline \end{array}$

2. $\begin{array}{r} 3,437 \\ \times \quad 20 \\ \hline \\ + \quad \\ \hline \end{array}$

3. $\begin{array}{r} 6,115 \\ \times \quad 25 \\ \hline \\ + \quad \\ \hline \end{array}$

4. $\begin{array}{r} 1,037 \\ \times \quad 31 \\ \hline \\ + \quad \\ \hline \end{array}$

5. $\begin{array}{r} 4,624 \\ \times \quad 29 \\ \hline \\ + \quad \\ \hline \end{array}$

6. $\begin{array}{r} 5,297 \\ \times \quad 33 \\ \hline \\ + \quad \\ \hline \end{array}$

7. $\begin{array}{r} 1,608 \\ \times \quad 37 \\ \hline \\ + \quad \\ \hline \end{array}$

8. $\begin{array}{r} 4,809 \\ \times \quad 47 \\ \hline \\ + \quad \\ \hline \end{array}$

9. $\begin{array}{r} 5,194 \\ \times \quad 10 \\ \hline \\ + \quad \\ \hline \end{array}$

10. $\begin{array}{r} 3,578 \\ \times \quad 36 \\ \hline \\ + \quad \\ \hline \end{array}$

11. $\begin{array}{r} 8,167 \\ \times \quad 57 \\ \hline \\ + \quad \\ \hline \end{array}$

12. $\begin{array}{r} 5,802 \\ \times \quad 66 \\ \hline \\ + \quad \\ \hline \end{array}$

DAY 5

MULTIPLYING 3-DIGIT NUMBERS BY 3-DIGIT NUMBERS

Solve the problems below.

1.
```
    3 6 7
  × 1 8 2
  -------
+
---------
```

2.
```
    2 2 7
  × 3 0 7
  -------
+
---------
```

3.
```
    5 2 1
  × 4 7 9
  -------
+
---------
```

4.
```
    1 3 9
  × 9 2 7
  -------
+
---------
```

5.
```
    8 6 5
  × 2 6 7
  -------
+
---------
```

6.
```
    5 5 6
  × 2 3 4
  -------
+
---------
```

7.
```
    6 8 2
  × 2 6 4
  -------
+
---------
```

8.
```
    9 0 5
  × 1 1 4
  -------
+
---------
```

9.
```
    9 4 6
  × 3 7 1
  -------
+
---------
```

Day 6

Multiplying 4-Digit Numbers by 3-Digit Numbers

Solve the problems below.

1.
$$\begin{array}{r} 6,584 \\ \times\ 324 \\ \hline \end{array}$$
+

2.
$$\begin{array}{r} 3,291 \\ \times\ 105 \\ \hline \end{array}$$
+

3.
$$\begin{array}{r} 8,347 \\ \times\ 137 \\ \hline \end{array}$$
+

4.
$$\begin{array}{r} 4,551 \\ \times\ 244 \\ \hline \end{array}$$
+

5.
$$\begin{array}{r} 9,067 \\ \times\ 320 \\ \hline \end{array}$$
+

6.
$$\begin{array}{r} 6,561 \\ \times\ 439 \\ \hline \end{array}$$
+

7.
$$\begin{array}{r} 7,812 \\ \times\ 140 \\ \hline \end{array}$$
+

8.
$$\begin{array}{r} 3,302 \\ \times\ 259 \\ \hline \end{array}$$
+

9.
$$\begin{array}{r} 5,191 \\ \times\ 341 \\ \hline \end{array}$$
+

Day 7

Dividing 5-Digit Dividends by 1-Digit Divisor

Find the quotients. Make sure to make any remainders into fractions and simplify.

1. $9\overline{)40{,}863}$

2. $6\overline{)92{,}025}$

3. $8\overline{)17{,}283}$

4. $7\overline{)59{,}518}$

5. $2\overline{)64{,}761}$

6. $5\overline{)51{,}909}$

7. $3\overline{)73{,}871}$

8. $8\overline{)81{,}095}$

9. $6\overline{)79{,}812}$

10. $4\overline{)33{,}299}$

11. $9\overline{)30{,}956}$

12. $7\overline{)99{,}068}$

Day 8

Division Word Problems

Use division to solve the problems below.

1. Reggie has 312 letters to deliver. If he delivers 6 letters an hour, how many hours will it take him to deliver all the letters?

____

2. Suzy is running a marathon. She needs to run 297 miles to train for the marathon. She has 11 days to train. How many miles per day does she need to run?

____

3. Ray has 714 square feet of snow to shovel from his yard. He can shovel 14 square feet an hour. How many hours will it take him to shovel all the snow?

____

4. Dad is building a fence. He has 595 boards. It takes 5 boards to build each section of the fence. How many sections of fence can he make with the boards he has?

____

Day 9

Dividing 5-Digit Dividends by a 2-Digit Divisor

Find the quotients. Make sure to make any remainders into fractions and simplify.

1. $29\overline{)66{,}126}$
2. $18\overline{)40{,}834}$
3. $43\overline{)54{,}957}$
4. $36\overline{)37{,}993}$
5. $56\overline{)76{,}013}$
6. $84\overline{)99{,}957}$
7. $70\overline{)32{,}142}$
8. $33\overline{)33{,}962}$
9. $44\overline{)59{,}195}$
10. $91\overline{)85{,}809}$
11. $62\overline{)65{,}135}$
12. $89\overline{)99{,}066}$

Day 10

Expanded Form to Exponent Form

Convert these to standard form of exponents.

1. $6 \times 6 \times 6 \times 6 =$ $\underline{\quad 6^4 \quad}$

2. $3 \times 3 \times 3 =$ ______

3. $5 \times 5 \times 5 =$ ______

4. $7 \times 7 \times 7 \times 7 =$ ______

5. $9 \times 9 \times 9 \times 9 \times 9 =$ ______

6. $8 \times 8 \times 8 \times 8 \times 8 \times 8 =$ ______

7. $2 \times 2 \times 2 \times 2 \times 2 \times 2 =$ ______

8. $4 \times 4 \times 4 \times 4 =$ ______

9. $5 \times 5 =$ ______

10. $6 \times 6 \times 6 \times 6 \times 6 =$ ______

11. $7 \times 7 \times 7 \times 7 \times 7 \times 7 \times 7 =$ ______

12. $2 \times 2 \times 2 =$ ______

13. $3 \times 3 \times 3 \times 3 =$ ______

14. $9 \times 9 \times 9 \times 9 \times 9 \times 9 =$ ______

Day 11

Exponent Form to Expanded Form to Standard Form

Write the exponents in expanded form and then put it in standard form by solving.

1. 9^3 = ____$9 \times 9 \times 9$____ = __729__
2. 5^5 = ____________ = ______
3. 8^4 = ____________ = ______
4. 2^9 ____________ = ______
5. 6^4 ____________ = ______
6. 4^5 ____________ = ______
7. 8^2 ____________ = ______
8. 6^5 ____________ = ______
9. 7^4 ____________ = ______
10. 3^8 ____________ = ______
11. 4^3 ____________ = ______
12. 5^7 ____________ = ______

DAY 12

COMPARING EXPONENTS

Compare the values using the symbols >, <, or =.

1. 2^4 [>] 3^2
2. 4^3 [ ] 3^2
3. 2^4 [ ] 5^2
4. 6^5 [ ] 4^3
5. 3^3 [ ] 5^2
6. 5^3 [ ] 7^2
7. 8^4 [ ] 5^3
8. 6^3 [ ] 9^3
9. 4^4 [ ] 8^4
10. 5^2 [ ] 9^2
11. 7^3 [ ] 2^2
12. 4^4 [ ] 3^2
13. 8^3 [ ] 4^2
14. 3^4 [ ] 9^3
15. 6^2 [ ] 7^2

Chapter 2: Integers and Operations

Integers are numbers that can be written without a fraction or a decimal. They can be positive, negative, or zero. Some examples include 7 and –56.

Negative numbers go to the left of zero on a number line, and positive numbers go to the right of zero on a number line. The further left on the number line, the smaller the value. The further right on the number line, the greater the value.

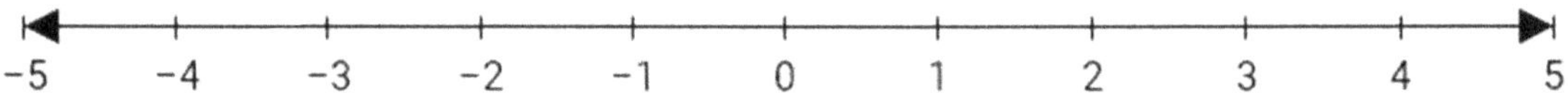

When you add two positive numbers, your answer will be positive. When you add two negative numbers, your answer will be negative. When you add a negative number to a positive number, the total value will be less than the positive number. Adding a negative number is the same as subtracting the positive version of that negative number.

$$\begin{aligned} &= 25 + (-5) \\ &= 25 - 5 \\ &= 20 \end{aligned}$$

When you add a positive number to a negative number, the total value will be greater than the negative number. When adding a positive number to a negative number, use subtraction.

$$\begin{aligned} &= -6 + 36 \\ &= 36 + -6 \\ &= 36 - 6 \\ &= 30 \end{aligned}$$

When you subtract two positive numbers, and the first number is larger, your answer will be positive. When you subtract two positive numbers, and the first number is smaller, your answer will be negative. When you subtract two negative numbers, your answer will be greater than the starting value. Subtracting a negative number is equivalent to adding the number.

$$-16 - (-4) = -16 + 4 = -12$$
$$-4 - (-16) = -4 + 16 = 12$$

When multiplying or dividing two positive numbers, the answer will be positive. When multiplying or dividing two negative numbers, the answer will be positive. When multiplying or dividing a negative number and a positive number, the answer will be negative.

$$-2 \cdot 8 = -16$$
$$24 \div -6 = -4$$

It is important to note that division can also be expressed as a fraction. For example, $24 \div -4 = -6$ could also be written as $\frac{24}{-4} = -6$.

A number's **opposite** is the number that is opposite to it on a number line or has an opposite sign. The opposite of –53 is 53. The opposite of 187 is –187. They both have the same distance from zero on a number line, just in different directions. The **absolute value** of a number is its distance from zero. Absolute value is notated by placing a vertical line on each side of a number. For instance, $|-46|$ means the absolute value of –46. Numbers and their opposite have the same distance from zero. –2 and 2 have a distance of 2 from zero. So, the absolute value of –2 is 2, or $|-2| = 2$. Also, the absolute value of 2 is 2, or $|2| = 2$. This may be noted by words, or with symbols.

Complete one worksheet each day in this chapter. Each day should take no longer than 15 minutes to complete.

Day 13

Integers on the Number Line

Place the following integers on the correct location on the number lines. Label the points with the letters.

a. 3 b. −5 c. −7 d. 2

e. −4 f. −1 g. −3 h. 1

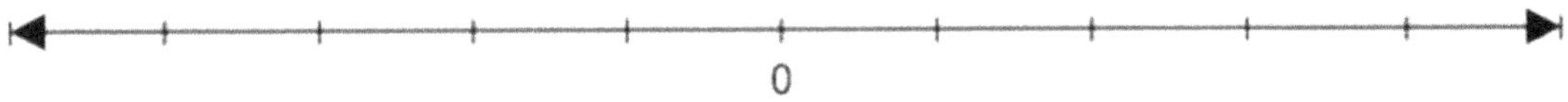

Day 14

Adding and Subtracting Integers

Use the number lines to add and subtract the following integer problems.

1.

$$4 + (-7) =$$

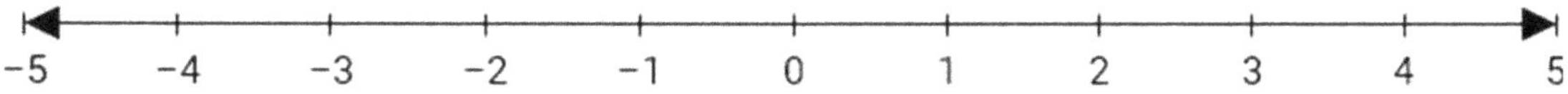

2.

$$-2 - 3 =$$

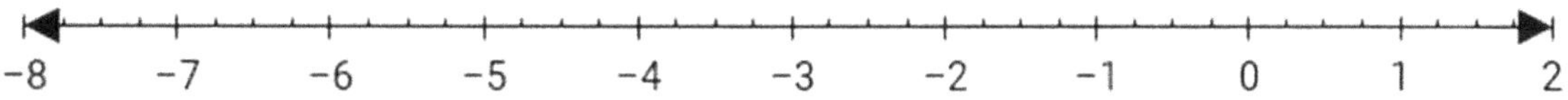

3.

$$1 + 3 =$$

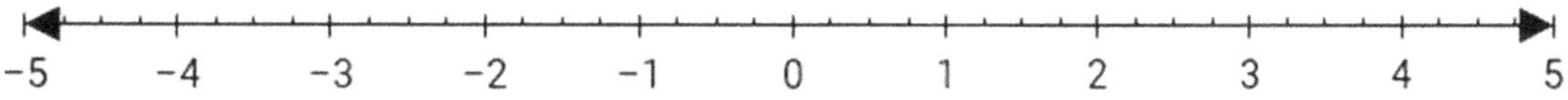

Solve the problems by adding or subtracting the integers.

4.

$$-13 + 6 =$$

5.

$$-6 + (\ 9) =$$

6.

$$6 - 7 =$$

7.

$$-10 + 3 =$$

8.

$$-2 - 5 =$$

Day 15

Multiplying and Dividing Integers

Solve the problems by multiplying or dividing the integers.

1. $-3 \times 7 =$

2. $6 \times -8 =$

3. $-4 \times -4 =$

4. $-16 \times 14 =$

5. $-90 \times 10 =$

6. $-224 \times -5 =$

7. $-42 \div -6 =$

8. $108 \div -9 =$

9. $200 \div -50 =$

10. $-144 \div -12 =$

11. $-468 \div -2 =$

12. $-196 \div 14 =$

Day 16

Absolute Value and Opposites

Determine the absolute value of the following problems.

1. $|-6|$

2. $|8|$

3. $|12|$

4. $|-13|$

5. $|66|$

6. $|-49|$

Determine the opposites of the following problems.

7. 5

8. 23

9. −17

10. −62

11. 45

12. −98

Chapter 3: Rational Numbers and Operations

Rational numbers can be positive or negative and include fractions, improper fractions, terminating decimals, and some repeating decimals. If it can be written as a fraction, it is rational. Some examples include 6, −23, $\frac{1}{4}$, and 0.46.

When multiplying decimals, you use the same strategy as multiplying whole numbers. The difference is that you have digits to the right of the decimal. When you are setting up a multiplication problem, it is helpful to place the number with more digits on the top, and the number with fewer digits will be on the bottom. Set up this problem traditionally, with no decimal places. The number 23,142 has more digits than 45, so 23,142 will be at the top of the problem, and 45 will be at the bottom.

$$
\begin{array}{r}
{\scriptstyle 1\;\;\;\;1\;\;\;\;} \\
{\scriptstyle \not{1}\;\;\;\;\not{2}\,\not{1}\;\;} \\
23{,}142 \\
\times \qquad 45 \\
\hline
115{,}710 \\
+\;\; 925{,}680 \\
\hline
1{,}041{,}390
\end{array}
$$

When using this strategy, add the decimal at the end. To figure out how many decimal places the answer will have, you count the number of digits behind the decimal point in each factor to find the decimal in the product. In this example, there are 3 decimal places in the first number (23.142) and 3 in the second number (0.045), so the answer will have 6 decimal places. Once the product is found, you would count back 6 digits and place the decimal in the appropriate place.

$$1.041390.$$

So, 23.142 times 0.045 is 1.04139, which can be rounded to the nearest thousandth to get 1.041.

Mixed numbers are values that are written using a whole number and a fraction. For example, $2\frac{1}{3}$ is a mixed number that consists of 7 groups of $\frac{1}{3}$. **Improper fractions** are fractions that have a numerator that is greater than the denominator. $\frac{4}{3}$ is an improper fraction that consists of 4 groups of $\frac{1}{3}$.

To convert a mixed number to an improper fraction, begin by multiplying the whole number by the denominator of the fraction. Next, add the product to the numerator of the fraction. This value becomes the numerator for the improper fraction. The denominator used in the mixed number will be the denominator used in the improper fraction. For example, $2\frac{3}{4} = \frac{11}{4}$.

To convert an improper fraction to a mixed number, divide the numerator by the denominator: the quotient is the integer part of the mixed number, and the remainder is the numerator. For example, $27 \div 8 = 3$ with a remainder of 3, so $\frac{27}{8} = 3\frac{3}{8}$.

Fractions are considered **equivalent** if they represent the same amount. You can make equivalent fractions through multiplication or division. For example, $\frac{6}{12} = \frac{18}{36}$. To achieve this, both the numerator and denominator of the starting fraction are multiplied by 3. Likewise, $\frac{18}{36} = \frac{6}{12}$. To achieve this, both the numerator and denominator of the starting fraction are divided by 3.

Simplifying fractions means reducing a fraction to the lowest possible equivalent term. To do this, find the greatest common factor (GCF) of the numerator and the denominator. Then divide both the numerator and denominator by that factor.

Fractions with the same denominator can be added or subtracted by adding or subtracting the numerators and keeping the denominators the same. Fractions with different denominators must first be made into an equivalent fraction so that the denominators are the same for each fraction involved in the addition or subtraction problem.

When you multiply fractions, multiply across the denominators and numerators. For example,

$$\frac{3}{1} \times \frac{5}{2} = \frac{3 \times 5}{1 \times 2} = \frac{15}{2} = 7\frac{1}{2}$$

To divide fractions, keep the first fraction the same, change the division sign to multiplication, and change the second fraction to the reciprocal. Then multiply like normal. This strategy is commonly referred to as "Keep, Change, Flip." To find the reciprocal of a fraction, switch the numerator and the denominator.

$$\frac{3}{4} \div \frac{1}{6} = \frac{3}{4} \times \frac{6}{1} = \frac{18}{4} = \frac{9}{2} = 4\frac{1}{2}$$

To convert a fraction to a decimal, you will divide the numerator by the denominator. For example, $\frac{2}{5}$ can be converted to a decimal by performing the division problem 2 divided by 5 to get the value 0.4, so 0.4 is equivalent to $\frac{2}{5}$.

$$\begin{array}{r} 0.4 \\ 5\overline{)2.0} \\ \underline{2\;0} \\ 0 \end{array}$$

$$\frac{2}{5} = 0.4$$

To convert a decimal to a percent, multiply the decimal by 100 and rewrite using the percent sign. For example, 0.4 can be converted to a percent by performing the calculation 0.4 times 100 to give the value 40, or 40%.

$$0.4 \times 100 = 40$$
$$0.4 = 40\%$$

To convert decimals to fractions, keep in mind place value and how a decimal is read. If you can remember the proper way to *read* a decimal, this will really help in generating a fraction that can then be simplified. If reading the decimal is difficult for you, you will need to identify each place value. For example, when you see a decimal like 0.20, you know that you can read this number as 20 hundredths, but also you can see that the smallest value is the zero in the hundredths place, so the fraction will be a part of 100. This can be represented as the fraction $\frac{20}{100}$, also read as 20 hundredths. Since 0.20 is equivalent to 0.2, you may also see this decimal as 2 tenths and represent the fraction $\frac{2}{10}$, also read as 2 tenths. Once you have this value, you can simplify by finding a common factor between the numerator and denominator to find $\frac{1}{5}$, so $0.20 = \frac{1}{5}$.

$$0.20 = \frac{20}{100} = \frac{2}{10} = \frac{1}{5}$$

To convert percentages to fractions, keep in mind that a percentage is a rate per 100. Any percent can be written as a part of 100. For example, 54% can be converted to a fraction by writing it as $\frac{54}{100}$. Now, this fraction can be simplified using common factors to find that $54\% = \frac{27}{50}$.

Complete one worksheet each day in this chapter. Each day should take no longer than 15 minutes to complete.

Day 17

Adding and Subtracting Fractions with Unlike Denominators

Add or subtract the fractions in the following problems. Be sure to put your answer in simplest form. Put your answers in the boxes.

1.

$$\frac{3}{4}+\frac{1}{7}$$

$$\frac{21}{28}+\frac{4}{28}=\frac{25}{28}$$

$$\frac{25}{28}$$

2.

$$\frac{12}{14}-\frac{3}{7}$$

3.

$$\frac{1}{10}+\frac{4}{20}$$

4.

$$\frac{6}{15}-\frac{8}{24}$$

5.

$$\frac{4}{18}+\frac{3}{9}$$

6.

$$\frac{8}{9}-\frac{4}{18}$$

7.

$$\frac{10}{35}+\frac{3}{5}$$

8.

$$\frac{19}{20}-\frac{4}{15}$$

9.

$$\frac{3}{11}+\frac{2}{3}$$

Day 18

Multiplying Fractions

Multiply the fractions in the following problems. Be sure to put your answer in simplest form. Put your answers in the boxes.

1. $\frac{4}{7} \times \frac{2}{8} = \frac{8}{56} = \frac{1}{7}$

   $\frac{1}{7}$

2. $\frac{3}{5} \times \frac{2}{3} =$

3. $\frac{15}{16} \times \frac{3}{4} =$

4. $\frac{6}{7} \times \frac{5}{9} =$

5. $\frac{13}{14} \times \frac{1}{3} =$

6. $\frac{2}{5} \times \frac{8}{9} =$

7. $\frac{2}{12} \times \frac{7}{8} =$

8. $\frac{11}{12} \times \frac{2}{7} =$

9. $\frac{9}{13} \times \frac{5}{6} =$

Day 19

Dividing Fractions

Divide the fractions in the following problems. Be sure to put your answer in simplest form. Put your answers in the boxes.

1.

$$\frac{3}{10} \div \frac{1}{3} =$$

$$\frac{3}{10} \times \frac{3}{1} = \frac{9}{10}$$

$$\frac{9}{10}$$

2.

$$\frac{1}{12} \div \frac{3}{4} =$$

3.

$$\frac{3}{8} \div \frac{1}{2} =$$

4.

$$\frac{2}{7} \div \frac{3}{5} =$$

5.

$$\frac{3}{5} \div \frac{9}{10} =$$

6.

$$\frac{1}{2} \div \frac{7}{8} =$$

Day 20

Converting Between Mixed Numbers and Improper Fractions

Convert the mixed numbers to improper fractions.

1. $2\frac{1}{3} = \frac{7}{3}$

2. $4\frac{1}{2} =$

3. $6\frac{2}{7} =$

4. $3\frac{8}{9} =$

5. $2\frac{1}{10} =$

6. $5\frac{3}{4} =$

Convert the improper fractions to mixed numbers.

7. $\frac{11}{4} = 2\frac{3}{4}$

8. $\frac{7}{6} =$

9. $\frac{24}{5} =$

10. $\frac{64}{7} =$

11. $\frac{33}{8} =$

12. $\frac{15}{2} =$

Day 21

Simplifying Fractions

Put these fractions in simplest form. Put your answers in the boxes.

1. $\frac{25 \div 5}{55 \div 5} = \frac{5}{11}$

   Answer: $\frac{5}{11}$

2. $\frac{36}{45}$

3. $\frac{22}{33}$

4. $\frac{42}{63}$

5. $\frac{56}{88}$

6. $\frac{27}{54}$

7. $\frac{18}{24}$

8. $\frac{80}{100}$

9. $\frac{108}{144}$

Day 22

Converting Fractions to Decimals and Percents

Convert these fractions to decimals, and then convert them to percents.

1. $\frac{3}{4} = 0.75 = 75\%$

2. $\frac{1}{4} = \qquad =$

3. $\frac{1}{8} = \qquad =$

4. $\frac{4}{10} = \qquad =$

5. $\frac{4}{5} = \qquad =$

6. $\frac{20}{100} = \qquad =$

7. $\frac{25}{50} = \qquad =$

8. $\frac{3}{5} = \qquad =$

Day 23

Converting Decimals to Fractions and Percents

Convert these decimals to fractions, and then convert them to percents.

1.

$0.30 = \frac{3}{10} = 30\%$

2.

$0.63 =$
$=$

3.

$0.50 =$
$=$

4.

$0.55 =$
$=$

5.

$0.25 =$
$=$

6.

$0.80 =$
$=$

7.

$0.375 =$
$=$

8.

$0.75 =$
$=$

Day 24

Converting Percents to Decimals and Fractions

Convert these percents to decimals, and then convert them to fractions.

1.

$88\% = 0.88 = \frac{22}{25}$

2.

$25\% =$
$=$

3.

$30\% =$
$=$

4.

$80\% =$
$=$

5.

$75\% =$
$=$

6.

$50\% =$
$=$

7.

$35\% =$
$=$

8.

$99\% =$
$=$

Day 25

Arranging and Decimal Operations

Line up the decimal values and add them together.

1. $164.99 + 805.46 + 0.75$

2. $150.92 + 321 + 50.24$

3. $95.56 + 128.7 + 9.23$

4. $389.04 + 84.99 + 6.15$

5. $18.56 + 991.5 + 525.09$

6. $339.1 - 0.799$

7. $53.210 - 37.059$

8. $14 - 7.095$

9. $579.802 - 0.07$

Day 26

Adding & Subtracting Decimals

Add the values to find the totals.

1.
2,418.09
4,334.49
1,482.11
+ 712.64

2.
6,967.24
5,110.37
2,326.22
+ 34.13

3.
2,126.94
3,615.03
4,331.62
+ 24.77

4.
7,269.81
1,382.43
409.30
+ 3.99

5.
7,291.39
3,036.13
9,123.24
+ 604.88

6.
4,168.68
5,055.33
4,333.07
+ 659.37

7.
120.589
- 100.075

8.
264.892
- 155.684

9.
49,078.2
- 38,865.1

10.
8,009.01
- 6,420.57

11.
9,569.25
- 3,365.72

12.
7.65940
- 2.29995

Day 27

Multiplying Decimals

Multiplying the decimal values below. Be sure to put the decimal point in the correct position in the final answer.

1.

```
   1 8. 3 2
×      6. 8
-----------

+
-----------
```

2.

```
   4 4. 0 9
×   0. 3 3
-----------

+
-----------
```

3.

```
   2 2. 3 9
×       7 4
-----------

+
-----------
```

4.

```
   5 7. 8 2
×      2. 9
-----------

+
-----------
```

5.

```
   7 7. 3 2
×   0. 5 5
-----------

+
-----------
```

6.

```
   3. 4 4 7
×      4. 1
-----------

+
-----------
```

7.

```
   5. 7 7 2
×       5 9
-----------

+
-----------
```

8.

```
   3 3. 4 7
×      3. 3
-----------

+
-----------
```

9.

```
   6 6 2. 5
×      1. 4
-----------

+
-----------
```

Day 28

Multiplying Decimals 2

Multiplying the decimal values below. Be sure to put the decimal point in the correct position in the final answer.

1. 239.4×3.92

+

2. 7.863×45.5

+

3. 53.22×8.45

+

4. 4.977×0.368

+

5. 369.9×15.4

+

6. 23.18×53.6

+

7. 946.2×0.337

+

8. 4.556×66.8

+

9. 327.4×4.19

+

Chapter 4: Expressions and Equations

The **order of operations** is a mathematical term to describe the correct sequence of steps to simplify an expression or solve an equation. The order of operations is very specific and important to remember. The acronym **PEMDAS** is used to help remember the order of operations.

The first operation is **parentheses**. If there are parentheses within other parentheses you should simplify from the inside out. The next operation is **exponents**. Again, if there are exponents inside other exponents, you should simplify from the inside out. The next group is **multiplication** and **division**. If multiplication and division are present, simplify from left to right. The last group is **addition** and **subtraction.** When there is more than one addition or subtraction operation, they are solved or simplified from left to right.

A **variable** is a letter used to represent an unknown numerical value in an expression or an equation.

An **expression** includes numbers, variables, or numbers and variables separated with operation signs. Anything that stands alone or could belong on one side of an equal sign, is an expression. $3x + 4$ is an example of an expression. It contains numbers, a variable, and an operation. A single number is also considered an expression. An **equation** will always contain an equal sign.

To simplify an expression, combine like terms and follow order of operations. **Like terms** are terms that have the same letter and same power. If the terms do not have a variable at all, they are like terms as well. If they match, you can combine them. For example, $3x + 2x$ simplifies to $5x$, and $6 + 2$ simplifies to 8. However, you cannot combine $3x + 6$, so it is already in simplest form. Furthermore, you cannot combine $5y + 3x$ since the variable letters are different. It is also in simplest form.

Given the equation $-6 + x = 10$, we can solve for x by performing operations to isolate x and get our variable on one side of the equation by itself.

In this example, we want to get x by itself on one side of the equation. We can do this by adding 6 to both sides of our equation.

$$\begin{aligned} -6 + x &= 10 \\ +6 &= +6 \\ x &= 16 \end{aligned}$$

We can do this because if we add, subtract, multiply, or divide the same value on both sides of the equation, the sentence will still be true since it is balanced. For instance, if $2 = 2$, then $2 + 3 = 2 + 3$.

$$\begin{aligned} 2 &= 2 \\ 2 + 3 &= 2 + 3 \end{aligned}$$

Now we know that the solution for $-6 + x = 10$ is $x = 16$. This means that if x equals 16, the sentence will be true.

The last strategy would be to use the values of x to test the solution. In this example, we have 16 as a possible solution, so we can use 16 in place of x to see if the equation is true.

$$\begin{aligned} -6 + x &= 10 \\ -6 + (16) &= 10 \\ 10 &= 10 \end{aligned}$$

When we simplify with 16 as the x-value, we get 10 on both sides of our equal sign, so our sentence is true and 16 is the solution.

Complete one worksheet each day in this chapter. Each day should take no longer than 15 minutes to complete.

Day 29

Determining the Variable

Read the word problems below. Determine what the variable or unknown value is in each problem. Do not actually answer/solve the problems.

1. Each shot of espresso that Grace gets from her favorite local coffee shop is $6. If Grace stops at the coffee shop 5 times a week, how much money will she spend at the shop each week?

____

2. Katherine spent $60 at the grocery store. She has a monthly budget of $450 dollars to spend. How much money does Katherine have left to spend?

____

3. Eric makes wood ornaments. He has a total of 14 orders to fulfill, but he has only made 6 of them. How many ornaments does he have left to make?

____

4. Katie has a goal to read 45 books this year. There are 52 weeks in a year and 365 days in a year. How many books does Katie have to finish per week to meet this goal?

____

5. Sam has a chocolate pie. He wants to share it with his family. He wants to make sure everyone has the same amount of pie, but he also wants to leave half of the pie for the next day. If Sam splits half the pie with 4 other people in his family, how much of the whole pie does each person get?

____

6. Nick wants to go on a trip to Spain in 14 months. It will cost him $4,350 to go on this trip. How much money does Nick need to save each month to be able to go on this trip?

____

Day 30

Expressions Versus Equations

Write whether each problem is an expression or an equation.

1. $6x + 42$

____

2. $3 + 6 - 7$

____

3. $21x = 63$

____

4. $56 - x = 35$

____

5. $12 + x - 15 + 21x$

____

6. $6(13x + 3)$

____

7. $28 - (24 \div 8) + 15$

____

8. $35x + (4 \times 13) = 122$

____

9. $64 = 8x$

____

10. $16 + 34x \div 2$

____

DAY 31

WRITING EXPRESSIONS AND EQUATIONS

Write out the expression or equation.

1. Twenty-six plus three times a number is equal to forty-four.

____

2. Eighteen divided by a number plus nine equals fifteen.

____

3. The product of thirty-one and a number, divided by two.

____

4. Twenty-four divided by three is eight.

____

5. Seventy-two divided by eight plus one.

____

6. Nine plus the difference of nineteen and the product of five and a number.

____

7. A number times the quotient of fifty-six and seven equals nine.

____

8. The difference of sixty-six and a number multiplied by ten.

____

Day 32

Simplifying Expressions

Put the following expressions in simplest form using order of operations and combining like terms. Be sure to circle or box your final answer.

1. $6 + 8 + 10$

2. $20 - 5 + 3$

3. $4 \times 3 + 6$

4. $36 \div 6 + 4$

5. $4 + 3x - 2x$

6. $2x + 5x$

7. $10y - 4y$

8. $6 + 14 \div 7$

9. $3(x + 4)$

10. $18 - 2z + 24 + 6z$

Day 33

Solving Equations

Solve the equations below. Be sure to circle or box your final answer.

1. $x + 5 = 12$

2. $y - 7 = 9$

3. $3x = 18$

4. $z \div 4 = 6$

5. $x + 8 = 20$

6. $2y = 14$

7. $a - 5 = 10$

8. $\frac{m}{5} = 3$

9. $4 + x = 11$

10. $6b = 36$

Chapter 5: Proportional Reasoning

A **ratio** expresses the relationship between two values. For example, comparisons of the amount of green paint to the amount of white paint, the number of dogs to number of cats, or the volume of flour to volume of sugar could be described using ratios. Ratios are expressed in a few ways: using the ratio symbol, using words, and as a fraction. For example, a veterinarian sees 35 dogs and 25 cats per week. The ratio of dogs to cats as 35 to 25, 35 : 25, or $\frac{35}{25}$. You can simplify a ratio by finding the greatest common factor and dividing each term which creates an equivalent ratio, such as $\frac{7}{5}$.

A **rate** is a comparison of two values with different units. Some common examples are the comparison of dollars to hour, miles to hours, and beats to seconds. Rates can be expressed in a variety of ways, including using language, as quotients, using equivalent rates, algebraically, using tables, or graphically. Let's say you get your first job working at a store making $12 for every hour of work. This is a rate, a comparison between two unlike units, dollars and hours. This rate can be expressed through words. We use the word *per* when expressing rates. You could say that you make 12 dollars per hour. Rates can be expressed as a fraction, or division. If you make 12 dollars per hour, you could express that as $12/hour, or $\frac{12\text{ dollars}}{1\text{ hour}}$. You could even say that the rate of dollars per hour is $\frac{12}{1}$.

A **unit rate** is a rate written in simplest form, where one of the two units is one. Let's say you work as a dog trainer for 4 hours and make $50. You are paid at a rate of $50 per 4 hours. The unit rate would describe how much you make in 1 hour.

To find the unit rate, this rate can be written as a fraction, $\frac{50}{4}$ dollars per hour. By simplifying the fraction, we find the unit rate.

$$\frac{50}{4} = \frac{12.5}{1} = 12.5$$

So, you work for $12.50 per hour.

We can represent ratios and rates using proportions and scale factors. A **proportion** is an equation with two equivalent ratios. We use proportions to compare ratios and to solve ratio and rate problems. Suppose a recipe requires 2 cups of sugar and 3 cups of flour. This can be described by the ratio 2 : 3, or $\frac{2}{3}$. Now, what if we were making a recipe with 8 cups of sugar? We can use proportions to help us solve this problem where x is the number of cups of flour needed.

$$\frac{2}{3} = \frac{8}{x}$$

Proportions are written in the form:

$$\frac{a}{b} = \frac{c}{d}$$

They can also be written in the form $ad = cb$ when the equation is rearranged.

$$\begin{aligned} d \times \frac{a}{b} &= \frac{c}{d} \times d \\ b\left(d \times \frac{a}{b}\right) &= c \times b \\ d \times a &= c \times b \\ ad &= cb \end{aligned}$$

This is also known as **cross multiplying**. When we have a proportion, we can cross multiply to solve.

$$\frac{a}{b} \times \frac{c}{d}$$

$$ad = cb$$

Now, solve the proportion for the number of cups of flour needed for 8 cups of sugar. So, if we are using 8 cups of sugar, we need 12 cups of flour.

$$\begin{aligned} \frac{2}{3} &= \frac{8}{x} \\ 2 \times x &= 8 \times 3 \\ \frac{2x}{2} &= \frac{24}{2} \\ x &= 12 \end{aligned}$$

These are equivalent ratios.

$$\frac{2}{3} = \frac{8}{12}$$

There are multiple different kinds of percent proportion problems. Let's look at a missing part problem. This would look like, "What is 50% of 20?"

In this problem, we need to find 50% of a whole value 20. We know that 50% of 20 is 10, but let's figure out how we can solve this mathematically, so we have a strategy to solve more difficult problems.

To find the value of the missing part, we multiply the whole by the percent in decimal form. The decimal form of 50% is 0.5, so we multiply 20 by 0.5 to find the value of the part.

$$20 \times 0.5 = 10$$

So, in a missing part problem, our equation is:

$$\text{whole} \times \text{percent as a decimal} = \text{part}$$

Complete one worksheet each day in this chapter. Each day should take no longer than 15 minutes to complete.

Day 34

Ratios

Read the scenarios. Write the ratios 3 different ways for each.

A bag has 3 red marbles, 4 purple marbles, 6 blue marbles, and 12 green marbles.

1. What is the ratio of red to blue?

2. What is the ratio of green to purple?

3. What is the ratio of purple to blue?

4. What is the ratio of green to red?

5. What is the ratio of purple to the total?

6. What is the ratio of red to the total?

There are 8 girls and 4 boys in a class.

7. What is the ratio of girls to boys?

8. What is the ratio of boys to girls?

9. What is the ratio of girls to the total?

10. What is the ratio of boys to the total?

11. Simplify the ratio girls to the total.

12. Simplify the ratio of boys to the total.

Day 35

Rates and Unit Rates

Read each problem and determine the rate and unit rate.

1. Deborah has 500 miles to travel. She can travel 120 in 4 hours using her car. What is the rate? What is the unit rate?

Rate: ________________________________

Unit Rate: ________________________________

2. Maria has $20. She buys 6 notebooks for a total of $12. What is the rate? What is the unit rate?

Rate: ________________________________

Unit Rate: ________________________________

3. James has a 462 page book to read. He reads 30 pages in 3 days. What is the rate? What is the unit rate?

Rate: ________________________________

Unit Rate: ________________________________

4. Samantha is training for a marathon. She jogs 2 miles in 20 minutes. What is the rate? What is the unit rate?

Rate: ________________________________

Unit Rate: ________________________________

5. John has $10 to buy juice at the store. He buys a pack of juice boxes for $8 that contains 8 juice boxes. What is the rate? What is the unit rate?

Rate: ________________________________

Unit Rate: ________________________________

Day 36

Proportions

Solve the following proportions. Show your work.

1.

$$\frac{2}{5} = \frac{x}{10}$$

2.

$$\frac{3}{4} = \frac{6}{x}$$

3.

$$\frac{5}{8} = \frac{x}{16}$$

4.

$$\frac{x}{7} = \frac{4}{14}$$

5.

$$\frac{6}{9} = \frac{x}{3}$$

Day 37

Percent Proportions

Solve the following proportions. Show your work.

1. What is 20% of 50?

$$\frac{x}{50} = \frac{20}{100}$$

2. What is 15% of 80?

$$\frac{x}{80} = \frac{15}{100}$$

3. A shirt costs $40. What is 25% of the price?

$$\frac{x}{40} = \frac{25}{100}$$

4. Find 10% of 120.

$$\frac{x}{120} = \frac{10}{100}$$

5. A backpack costs $60. What is 30% of the price?

$$\frac{x}{60} = \frac{30}{100}$$

Chapter 6: Geometry

The **area** of a shape describes the total measure of the amount of space inside the polygon. We describe area using **square units** because area can be described as the number of squares that cover the inside of a shape.

We can also use formulas to find the area of a rectangle. Instead of counting the square units, we can find the area of a rectangle by multiplying the base and the height.

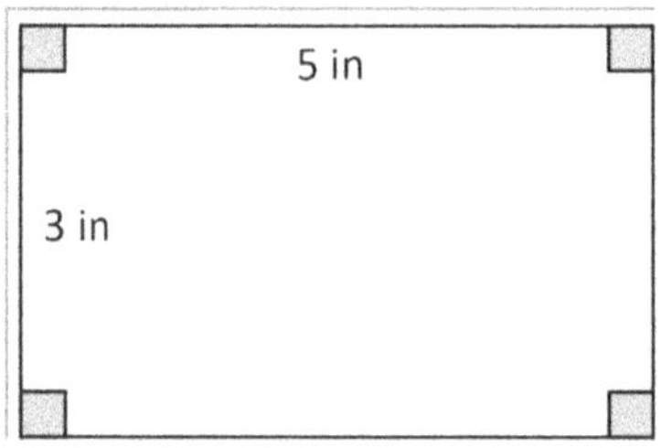

$$A = 5 \text{ in} \times 3 \text{ in}$$
$$A = 15 \text{ in}^2$$

A triangle is half of a rectangle with the same base and height, we can determine that the area of a triangle can be found by multiplying the base and height of the triangle and dividing this product by 2.

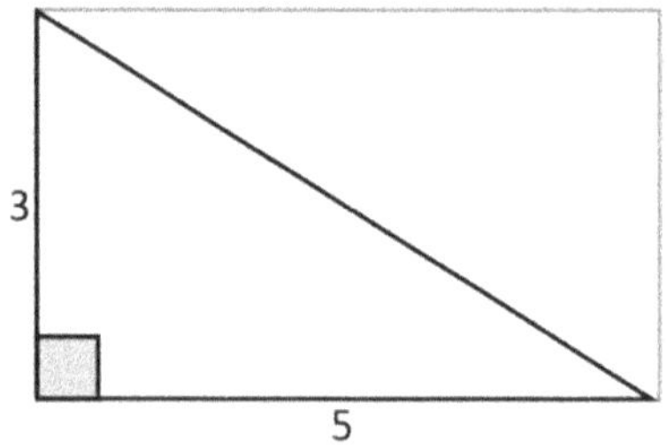

$$A = \frac{1}{2}(3 \times 5)$$
$$A = 7.5 \text{ units}^2$$

The area of a polygon is the same as the sum of its parts, so we can use what we know about rectangles and triangles to find the area of the parallelogram. A parallelogram has two sets of parallel sides. This means that a parallelogram has two identical triangles on either side. If we rearrange these "triangle ends" we can see that a rectangle with the same measurements is formed. We know that the area of a rectangle can be found using the formula $A = b \cdot h$, which is the same formula as a parallelogram. The area of a parallelogram is equivalent to the area of a rectangle with the same base and height measurements.

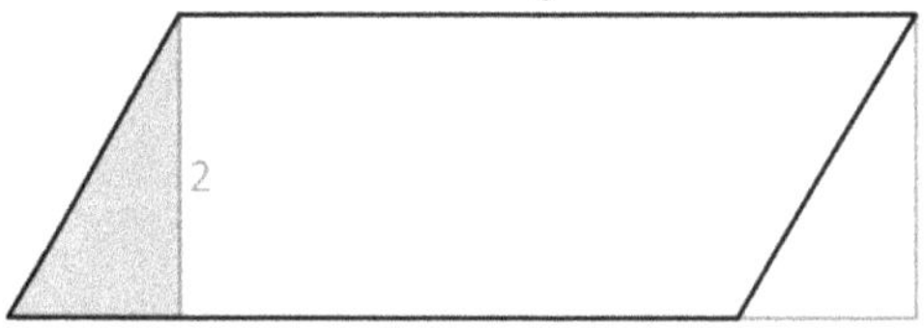

$$A = 5 \times 2$$
$$A = 10 \text{ units}^2$$

A **trapezoid** can be viewed as part of a parallelogram. If you make a second copy and rotate it next to the original, a parallelogram is formed. A trapezoid has two bases because there are two different sides that are perpendicular to the height. The height of the parallelogram is h and the base of the parallelogram is $b_1 + b_2$. The area of the parallelogram is $(b_1 + b_2) \times h$. To find the area of the trapezoid, we need to divide the big parallelogram area by 2, or multiply it by $\frac{1}{2}$. This gives the formula: $A = \frac{1}{2}(b_1 + b_2) \times h$.

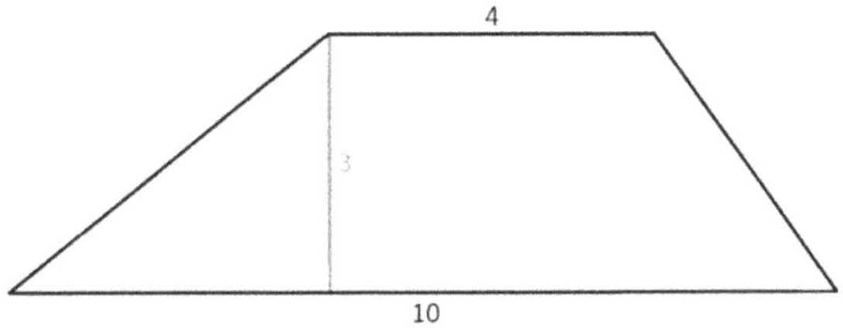

$$A = \frac{1}{2}(b_1 + b_2) \times h$$
$$A = \frac{1}{2}(4 + 10) \times 3$$
$$A = \frac{1}{2}(14 \times 3)$$
$$A = \frac{1}{2}(42)$$
$$A = 21 \text{ units}^2$$

Volume is the measurement of space inside a three-dimensional shape or object. We use **cubic units**, which can be thought of as cubed units, to describe volume because volume is the number of unit cubes it takes to fill a space. A **rectangular prism** is a three-dimensional shape that has 6 rectangular faces. To find the volume of a rectangular prism, we find the area of a face and multiply that by the height of the prism. In other words, we find the volume of the first layer, and multiply that by how many layers make up the prism.

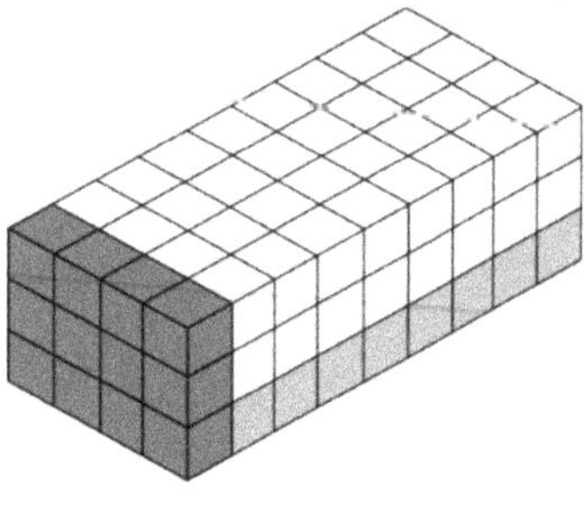

More simply put, the volume of a rectangular prism is the length times the width times the height, or $V = l \cdot w \cdot h$.

Complete one worksheet each day in this chapter. Each day should take no longer than 15 minutes to complete.

Day 38

Finding Area of Triangles

Find the area of each shape. Write the whole problem out. Do not forget to include the units in your final answer.

1.

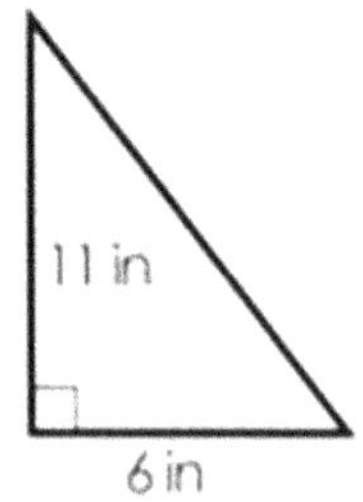

2.

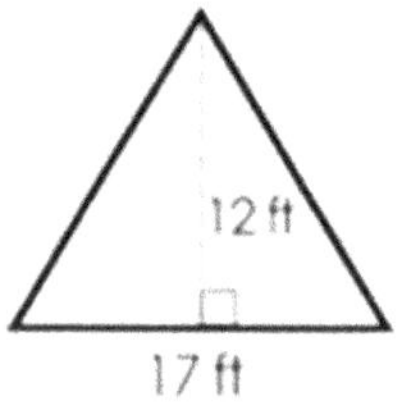

3.

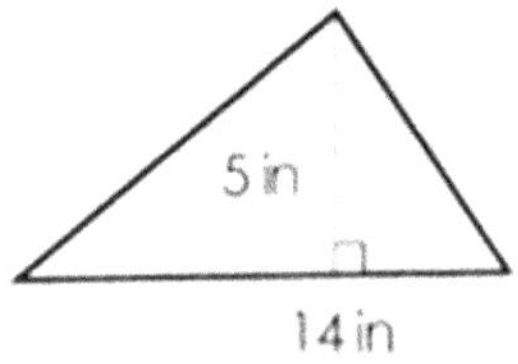

4.

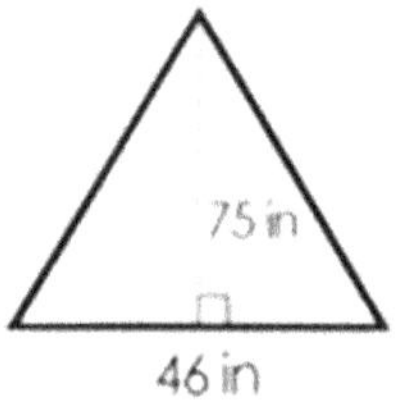

5.

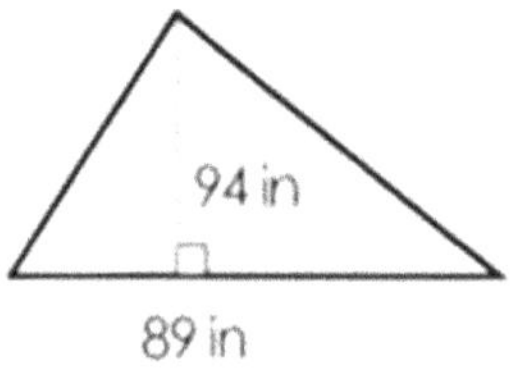

6.

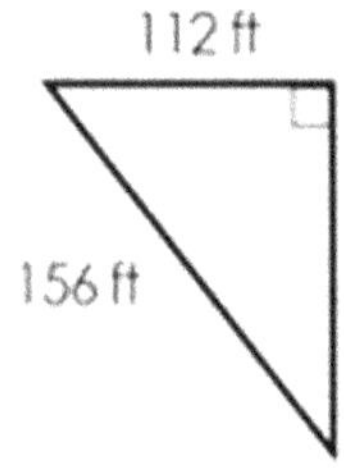

DAY 39

FINDING AREA OF PARALLELOGRAMS

Find the area of each shape. Write the whole problem out. Do not forget to include the units in your final answer.

1.

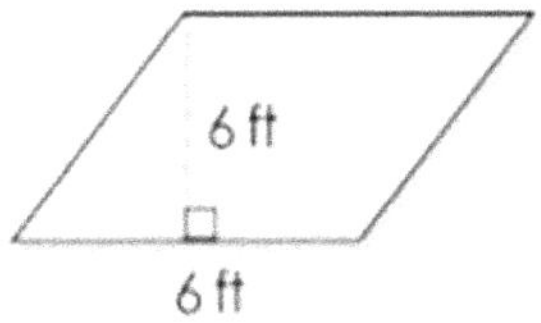

2.

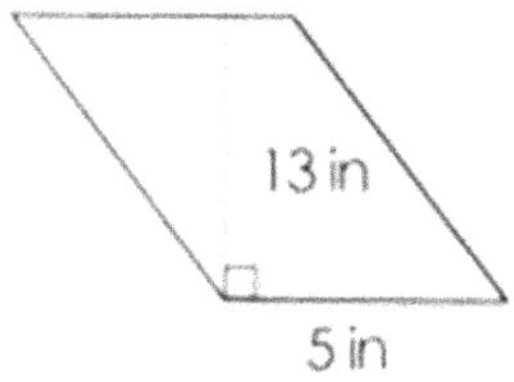

3.

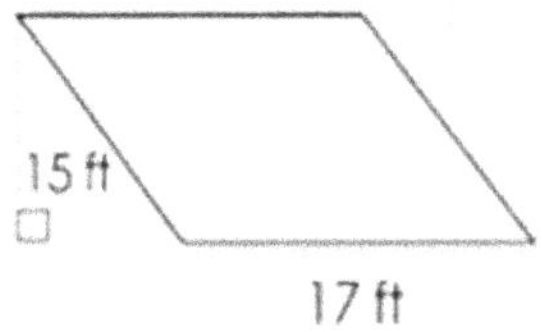

4.

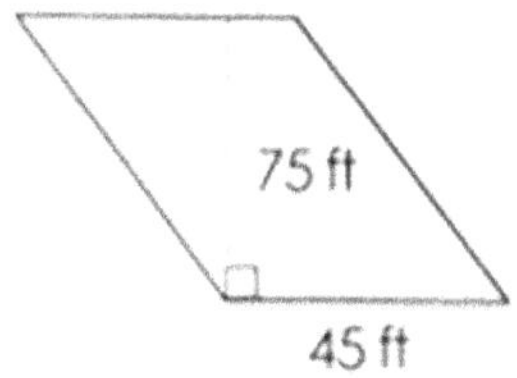

5.

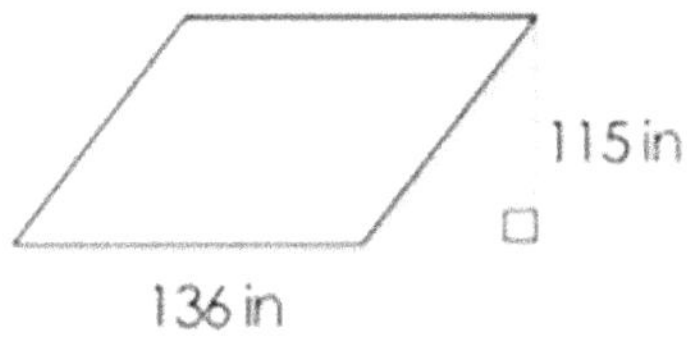

6.

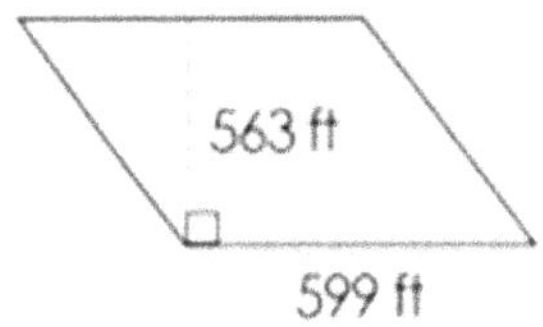

Day 40

Finding Area of Trapezoids

Find the area of each shape. Write the whole problem out. Do not forget to include the units in your final answer.

1.

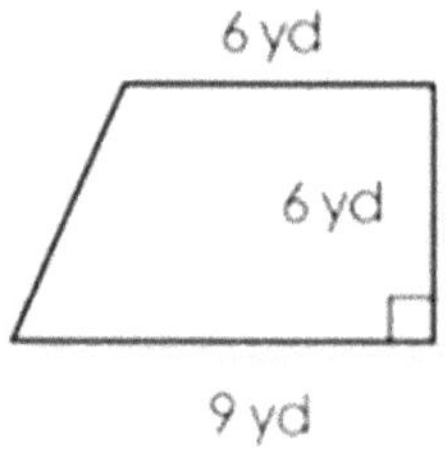

____

2.

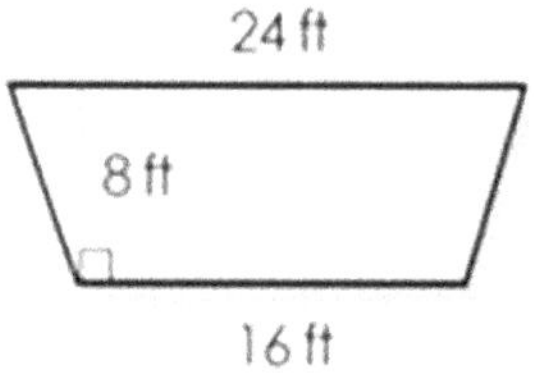

____

3.

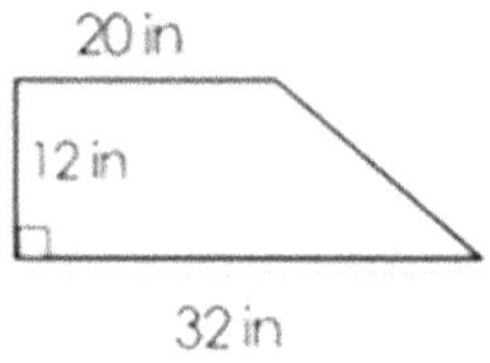

____

4.

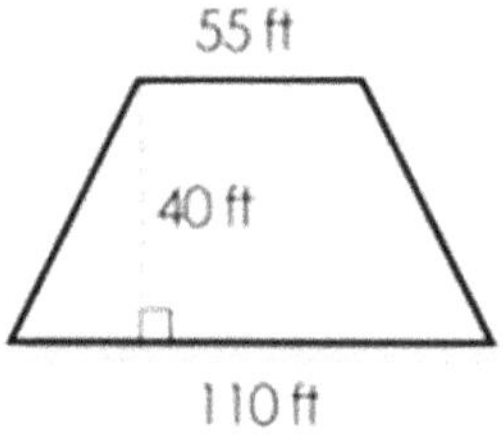

____

5.

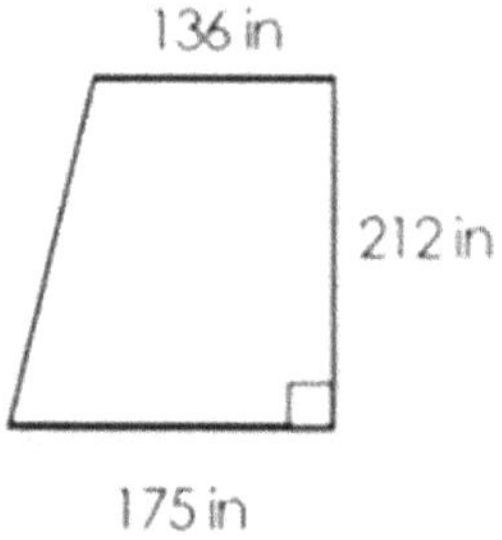

____

6.

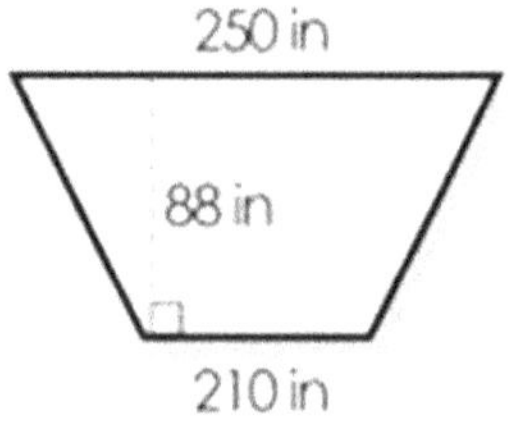

____

Day 41

Finding Perimeter

Find the perimeter of each shape. Write the whole problem out. Do not forget to include the units in your final answer.

1.

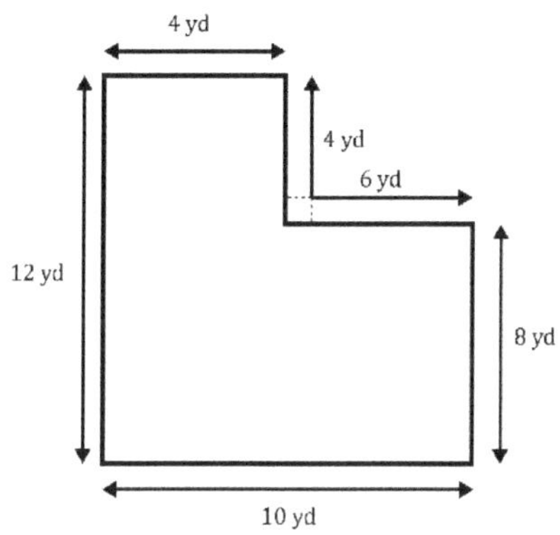

2.

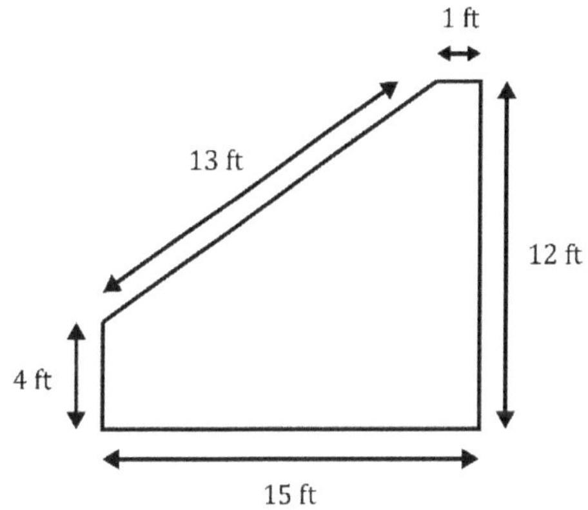

3.

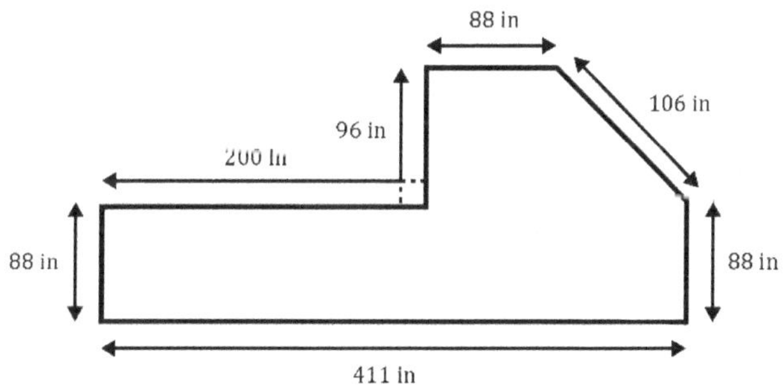

4.

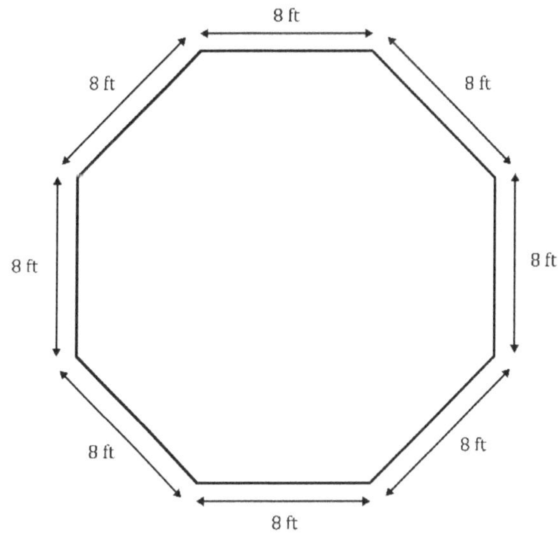

5.

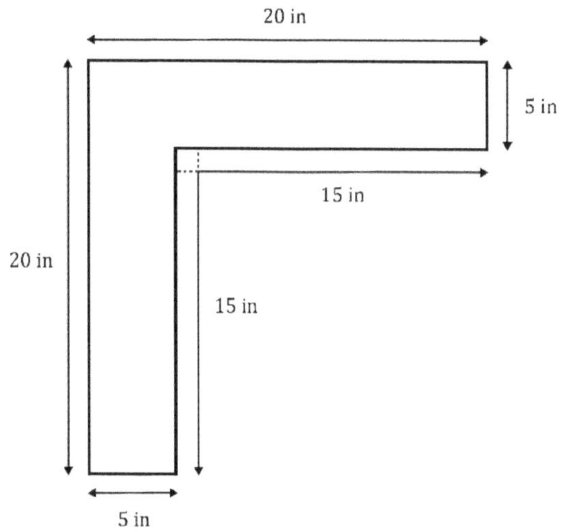
20 in
5 in
15 in
20 in
15 in
5 in

6.

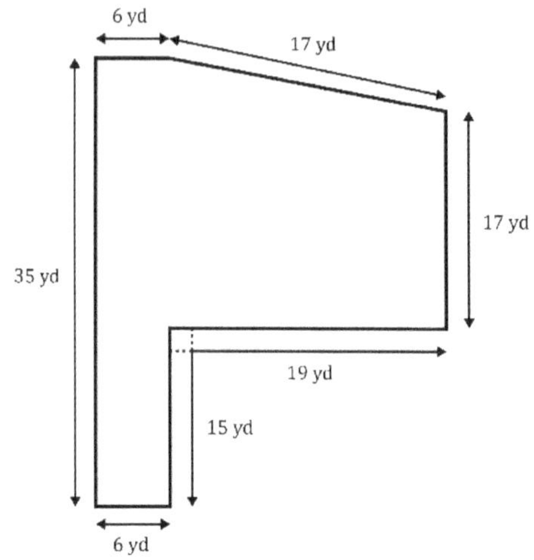
6 yd
17 yd
17 yd
35 yd
19 yd
15 yd
6 yd

7.

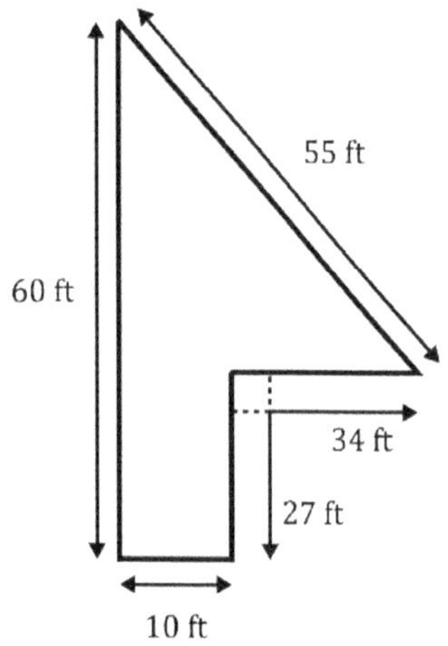
55 ft
60 ft
34 ft
27 ft
10 ft

8.

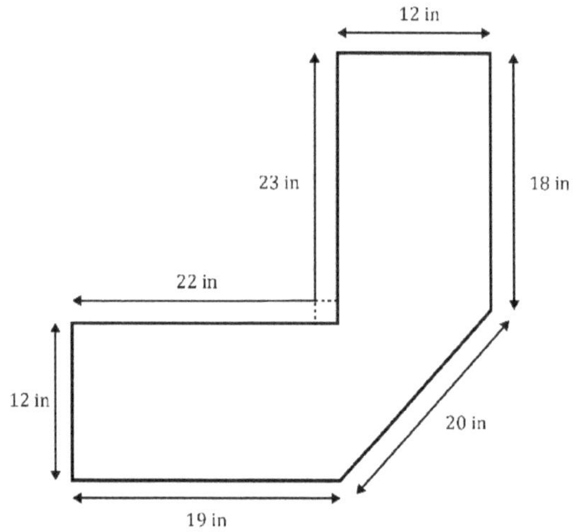
12 in
23 in
18 in
22 in
12 in
20 in
19 in

Day 42

Volume of Rectangular Prisms

Find the volume of the following rectangular prisms.

1. A rectangular pool is 30 feet by 20 feet by 6 feet. What is the volume?

____

2. Carrie has a rectangular vase that is 5 inches by 5 inches by 12 inches. What is the volume?

____

3. What is the volume of the rectangular prism below?

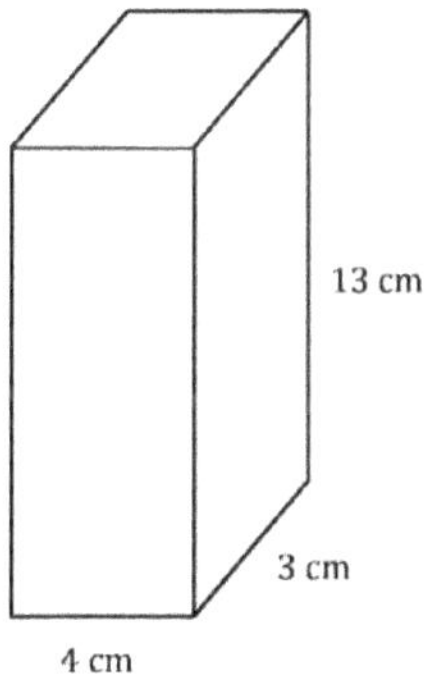

____

4. What is the volume of the rectangular prism below?

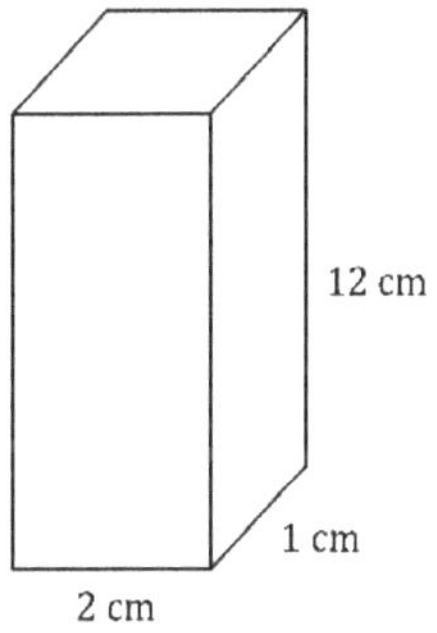

____

5. Logan has a piece of wood for a post that is 4 inches by 4 inches by 72 inches. What is the volume?

____

6. A rectangular raised garden bed is 6 feet by 4 feet by 2 feet. What is the volume?

____

7. What is the volume of the rectangular prism below?

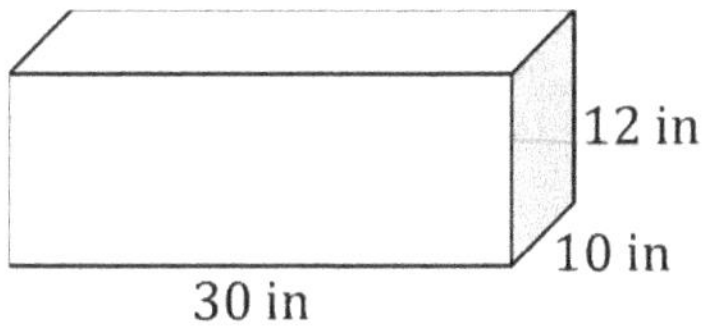

____

8. What is the volume of the rectangular prism below?

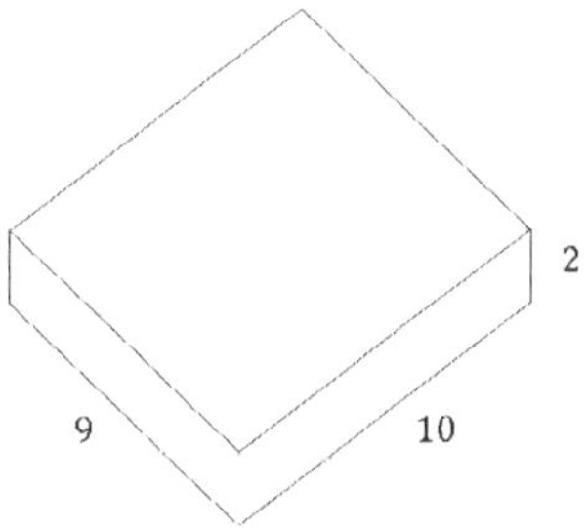

____

Chapter 7: Graphing & Data Analysis

Data refers to a collection of information. Data can be summarized using different graphical displays such as **bar graphs, histograms, pie charts, stem-and-leaf plots,** and **box-and-whisker plots**.

Bar graphs use horizontal or vertical bars to represent the total amount for each category. For example, this bar graph shows the number of votes for a variety of snacks.

A **histogram** organizes data by showing the frequency of data in certain ranges. A histogram contains a vertical and horizontal axis. Typically, the frequency is listed vertically. Histograms can illustrate many types of data sets. It looks similar to a bar graph, but the bars are touching.

A **pie chart**, also known as a circle graph, is a type of graph that uses a circle to represent the total data and each portion of the circle is divided based on the way the data is distributed. Pie charts are commonly used with percentages and to illustrate how something is distributed across groups. A complete circle of the pie chart is 100%, and the "slices" represent different portions.

A **stem-and-leaf plot** is a way to illustrate data with multi-digit values. This type of data representation is best used with a large number of data points that vary in value, say values from 0 to 100. In a stem-and-leaf plot, values are separated by place value. There is a list of numbers listed vertically along the left side; these are the stems. There is a list of numbers to the right of each stem; these are the leaves. There is a line dividing the stems from the leaves.

Stem	Leaf
0	2, 2, 4, 5, 8, 8, 9
1	1, 2, 3, 3, 3, 4, 4, 5, 7, 9
2	0, 2, 8
3	2, 4, 6, 6, 7
4	3, 4, 9
5	0, 4, 5, 5, 6
6	4, 5, 5
7	0, 1
8	2
9	0, 4
10	2

Key:

9|4 = 94

10|2 = 102

Box plots, or **box-and-whisker plots,** show the minimum value, the maximum value, the median value, the first quartile, and the third quartile of a data set. The minimum value is the smallest value and the maximum value is the largest value. The median is the exact center of a data set.

The **first quartile** is the median of the lower half of the data set, or the median of the values to the left of the median. The **third quartile** is the median of the upper half of the data set, or the median of the values to the right of the median.

This list gives the heights in inches of students in a 6th grade math class.

{52, 52, 53, 55, 56, 56, 57, 57, 57, 57, 58, 58, 59, 59, 59, 59, 60, 60, 62, 62, 64, 64, 65, 65}

This box plot represents the data in this list.

Minimum Value | First Quartile | Median | Third Quartile | Maximum Value

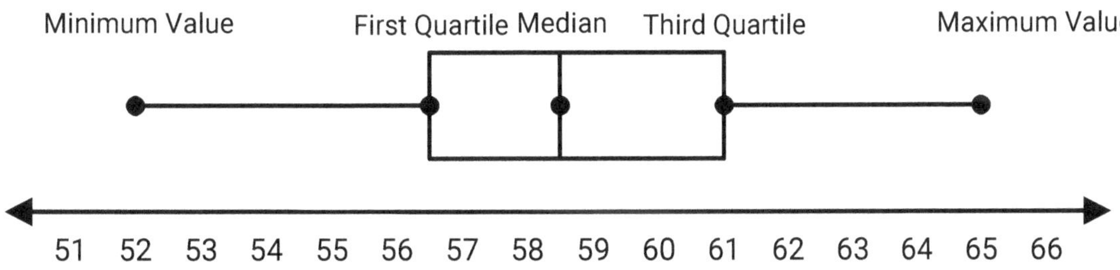

The point farthest to the left represents the minimum value in the data set, 52 inches.

The point farthest to the right represents the maximum value in the data set, 65 inches.

The point inside the box represents the median value in the data set, 58.5 inches, the mean of 58 and 59.

The first quartile is represented by the edge of the box to the left of the median. This value is the median of the values to the left of the median, 56 inches.

The third quartile is represented by the edge of the box to the right of the median. This value is the median of the values to the right of the median, 62 inches.

A box plot represents the four quarters of a data set.

The **measures of center** are different methods for estimating or measuring the center of a data set. The center of a data set can be described by the most frequent value in a data set, the value appearing in the middle of a data set arranged from least to greatest, or the value that is the average of that data set. The terms used to describe the center of a data set are mode, median, and mean.

The **mode** of a data set is the number that shows up most frequently.

Let's say we are given this data set.

{0, 0, 1, 1, 1, 1, 1, 1, 1, 1, 1, 1, 2, 2, 2, 2, 3, 4, 5}

The mode in a data set is the number that repeats the most frequently. If we look at this list of values, we can see that 1 shows up the most often, so the mode of this data set is 1.

The **median** of a data set is the literal middle value in a set of values. If you have a list of values, the median is going to be the value directly in the middle of a list of these values when the values are listed from least to greatest.

Let's say we have a set of data in a list.

$$\{3, 2, 2, 3, 0, 1, 2, 1, 1, 6, 2, 4, 2, 3, 6, 0, 0, 1, 1, 2, 4, 2, 4, 3, 1\}$$

This is a list of values, and we can determine the center, or median value in the data set by lining the values up from least to greatest and finding the middle value.

$$\{0, 0, 0, 1, 1, 1, 1, 1, 1, 2, 2, 2, 2, 2, 2, 2, 3, 3, 3, 3, 4, 4, 4, 6, 6\}$$

Now that we have our data set listed from least to greatest, the median will be the middle number in the set.

$$\overbrace{0\ 0\ 0\ 1\ 1\ 1\ 1\ 1\ 1\ 2\ 2\ 2}^{12}\ \textcircled{2}\ \overbrace{2\ 2\ 2\ 3\ 3\ 3\ 3\ 4\ 4\ 4\ 6\ 6}^{12}$$

The center, or median value in this data set is 2, because when you list the values from least to greatest, 2 is the middle number. It is in the middle because it is the same distance from the smallest value to the greatest value and there are the same number of values, 12, on each side.

Finding the median of a data set that has even number of values requires a slightly different approach.

Let's say we are given this data set.

$$\{15, 16, 18, 20, 22, 12, 13, 14, 19, 21, 21, 11, 12, 14\}$$

First, we need to list the values from least to greatest to find the middle.

$$\{11, 12, 12, 13, 14, 14, 15, 16, 18, 19, 20, 21, 21, 22\}$$

Now, we can count to find the middle.

11 12 12 13 14 14 15|16 18 19 20 21 21 22

In this data set, there is not a middle number. Instead, the middle position lies between 15 and 16. So, to find the median value, we will have to find the mean of the two middle numbers, 15 and 16.

To find the mean of these middle values, we will add them together and divide by 2.

$$15 + 16 = 31$$
$$31 \div 2 = 15.5$$

So, the median of this data set is 15.5.

The **mean** of a data set is often called the **average**. To find the mean of a data set, add the values in the data set together, and divide the sum by the total number of values.

In school, we are familiar with the mean of data sets because this is the way average grades are calculated. To get a final grade, often teachers will add the grade on each assignment or test together and divide by the total number of assignments or tests.

Let's say you are in a math class and your scores are listed in a data set.

$$\{76, 82, 82, 84, 87, 90, 85, 100, 94, 90\}$$

To find the mean, or average, of your scores, add the values and divide by the number of values. In this data set, if we add all the given grades together, we will get a total of 870. Now, if we divide 870 by the total number of grades, 10, we will get 87. That means that the final grade, the average grade, or the mean of the data set is 87.

The **spread** of a data distribution is a way to describe the range, or how spread out or close together the data is. The range of data is the difference between the smallest value and the largest value.

A **coordinate plane** is a way to graph data by plotting points on a grid. On a coordinate plane, there is an x-axis and a y-axis. The center point is called the **origin**, and its coordinates are (0,0). The x-axis runs horizontally, and the y-axis runs vertically. We use the x- and y-axis to graph ordered pairs. **Ordered pairs** are sets of values that tell us the coordinates of a point. They are written using parentheses and separated by a comma. (3,2) is an example of an ordered pair. In an ordered pair, the first number is the x-value, and the second number is the y-value. In this example, (3,2), $x = 3$ and $y = 2$. We use these values to plot points on a graph. The point (3,2) lies three units to the right of the origin along the x-axis and two units up along the y-axis. The point also lies at the intersection point of the lines $x = 3$ and $y = 2$. Positive values always run to the right along the x-axis and up along the y-axis. Negative values always run to the left along the x-axis and down along the y-axis.

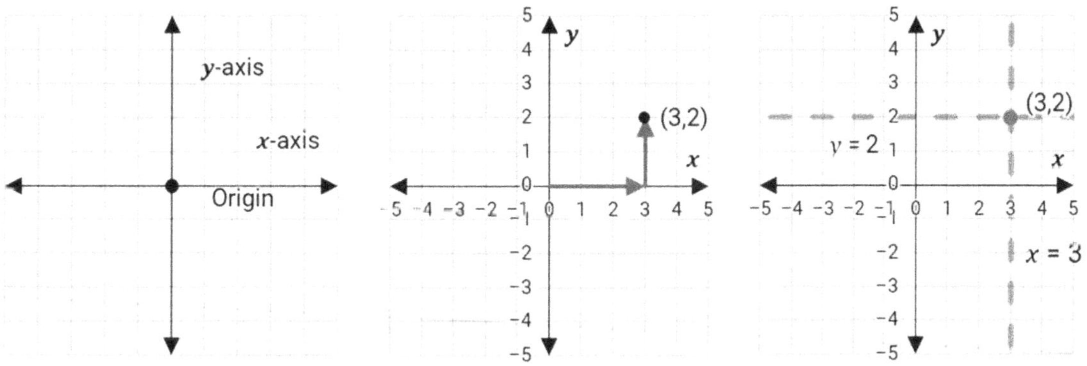

Complete one worksheet each day in this chapter. Each day should take no longer than 15 minutes to complete.

DAY 43
BAR GRAPHS

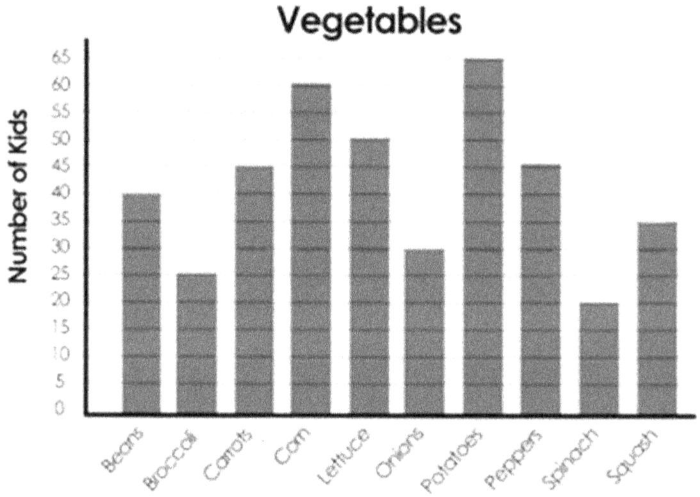

Use the bar graph above to answer the questions below.

1. How many kids eat corn? ________

2. How many kids eat squash? ________

3. How many kids eat carrots? ________

4. How many kids eat onions? ________

5. How many kids eat peppers? ________

6. How many kids eat lettuce? ________

7. How many kids eat potatoes? ________

8. How many kids eat broccoli? ________

9. How many kids eat beans? ________

10. How many kids eat spinach? ________

11. How many kids prefer corn over squash? ________

12. How many kids prefer broccoli over spinach? ________

13. How many kids prefer potatoes over beans?

14. How many kids prefer lettuce over peppers? ________

15. How many kids prefer carrots over broccoli? ________

16. How many kids prefer corn over onions? ________

17. How many kids prefer potatoes over spinach? ________

18. How many kids prefer beans over onions? ________

Day 44
Histograms

Use the histogram below to answer the questions that follow.

The 6th grade teachers at Washington Elementary School are doing a collaborative unit on cherry trees. Miss Wilson's math classes are making histograms summarizing the heights of black cherry trees located at a local fruit orchard.

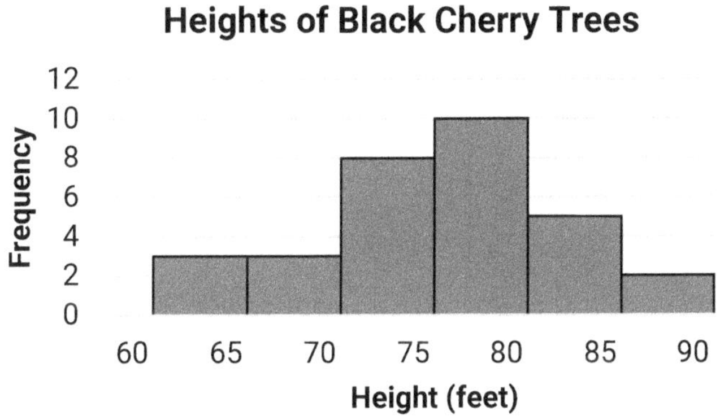

1. What is being measured in this histogram?

2. What does each bar in the histogram show?

3. Which height range has the greatest number of black cherry trees?

4. About how many trees are between 75 and 80 feet tall?

5. Are there more trees that are shorter than 75 feet or taller than 75 feet?

6. What is one conclusion Miss Wilson's class could make about the heights of black cherry trees at the orchard based on the histogram?

DAY 45

PIE CHARTS

Show Types that Family Members Prefer

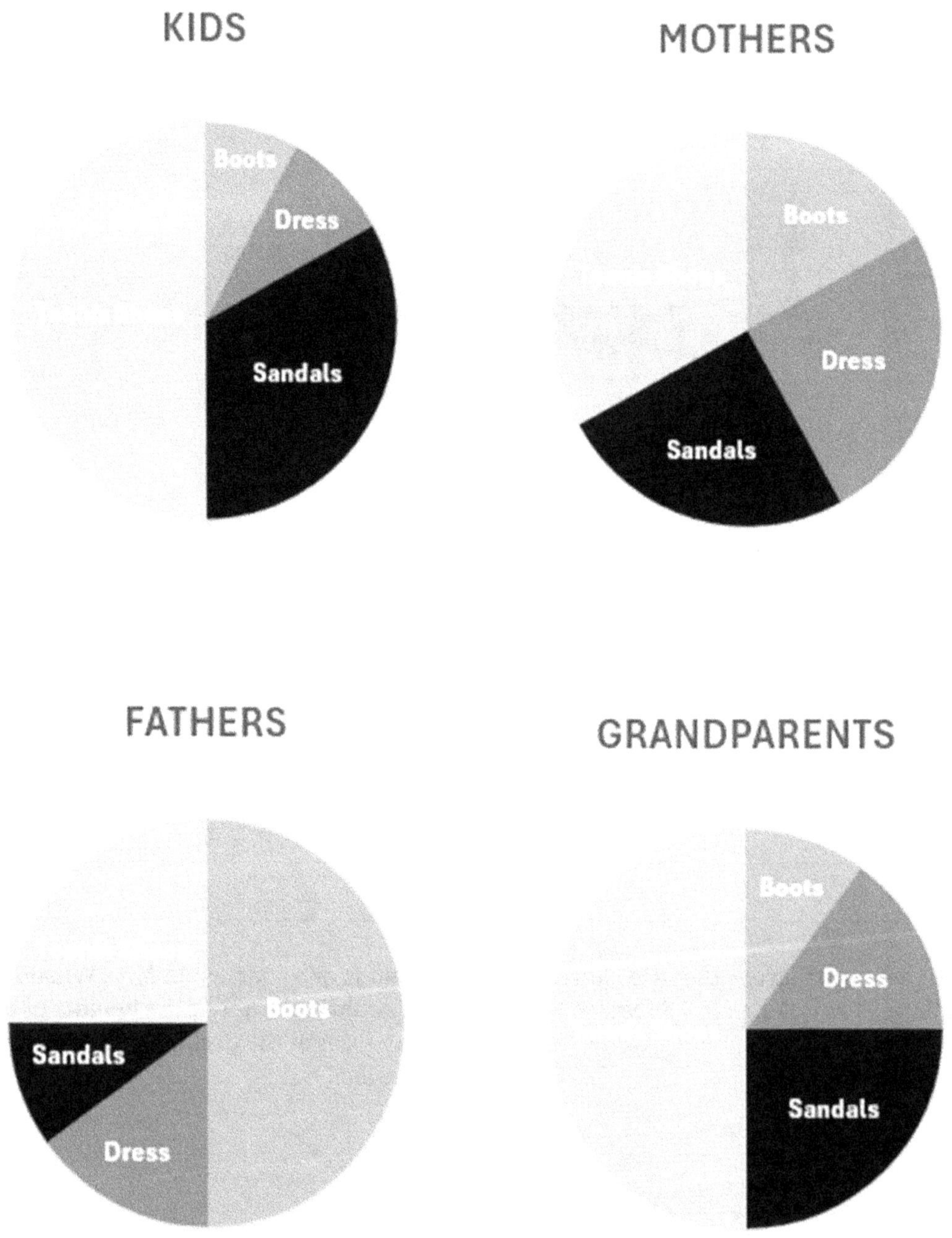

Use the pie charts on the previous page to answer the questions below.

1. Which shoes do kids like the most? ______________________

2. Which shoes do grandparents like the most? ______________________

3. Which shoes do mothers like the most? ______________________

4. Which shoes do fathers like the most? ______________________

5. What kind of shoe is second favorite for kids? ______________________

6. What kind of shoe is second favorite for fathers? ______________________

7. What shoe do kids like the least? ______________________

8. What shoe do grandparents like the least? ______________________

9. What shoe do mothers like the least? ______________________

10. What shoe do fathers like the least? ______________________

Day 46

Stem & Leaf Plots

Use the stem and leaf plot to answer the questions that follow.

The stem and leaf plot shows final exam scores for a statistics class.

Stem	Leaf
5	6
6	7 7 9
7	2 4 7 7 8
8	1 2 2 3 4 8
9	0 2 3 4

Key: 5|6 = 56

1. How many students took the final exam?

____

2. What is the lowest score on the final exam?

____

3. What is the highest score on the final exam?

____

4. How many students scored in the 70s?

____

5. What score appears most often in the data?

____

Day 47

Box and Whisker Plots

Use the information below to construct a boxplot, and then answer the questions that follow.

An Algebra 1 class recently took a test on solving equations. The student scores are listed below:

80, 90, 72, 74, 100, 77, 77, 78, 90, 92, 95, 98, 100, 80, 82, 88

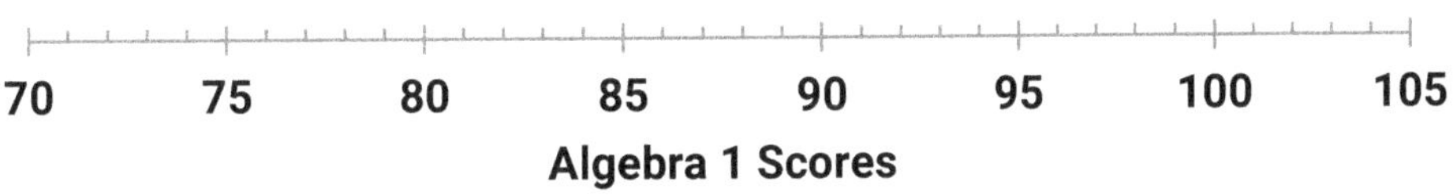

1. What is the minimum value? ______________

2. What is the maximum value? ______________

3. What is the median value? ______________

4. What is the first quartile value? ______________

5. What is the third quartile value? ______________

Day 48

Mean, Median, Mode, and Range

Find the mean, median, mode, and range of each of the following data sets.

1.

80, 90, 91, 24, 36, 99, 90, 92, 56, 79, 70, 70, 80, 79, 79, 44, 95, 90, 79,77

Mean: ___________ Median: ___________ Mode: ___________ Range: ___________

2.

5, 6, 5, 7, 4, 9, 5, 7, 8, 9, 9, 5, 7, 4, 3, 2, 5, 8, 2, 9, 9

Mean: ___________ Median: ___________ Mode: ___________ Range: ___________

3.

24, 26, 27, 24, 24, 27, 31, 25, 21, 26, 28, 29, 26

Mean: ___________ Median: ___________ Mode: ___________ Range: ___________

DAY 49

GRAPHING ON A COORDINATE PLANE

Use the coordinate planes to graph the ordered pairs.

1.

$W = (-2,5)$

$X = (-2,-5)$

$Y = (5,-2)$

$Z = (-5,2)$

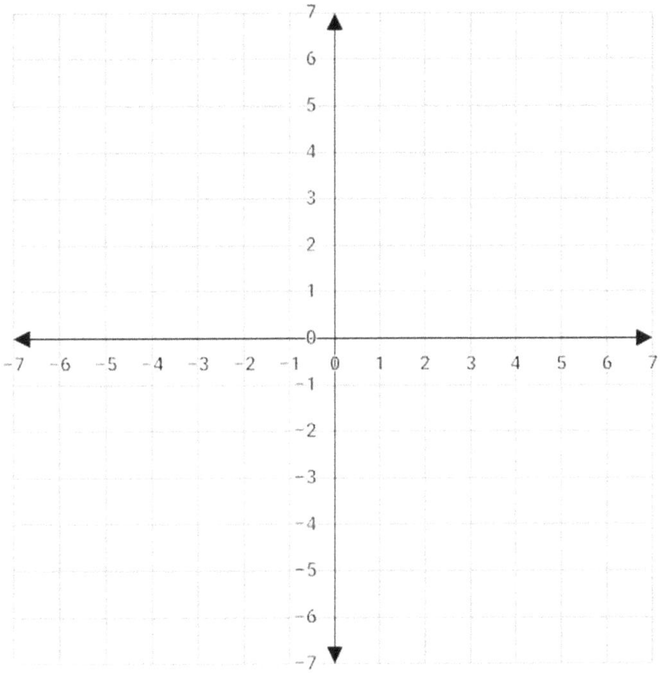

2.

$R = (-3,2)$

$S = (-1,3)$

$T = (3, \quad 6)$

$U = (-5,-5)$

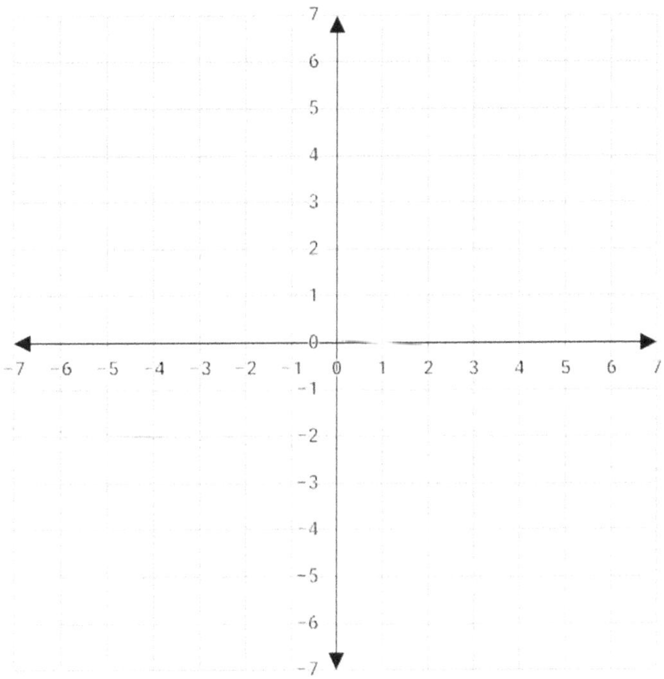

DAY 50

DETERMINING ORDERED PAIRS USING A COORDINATE PLANE

Use the coordinate plane below to determine the ordered pairs for each point.

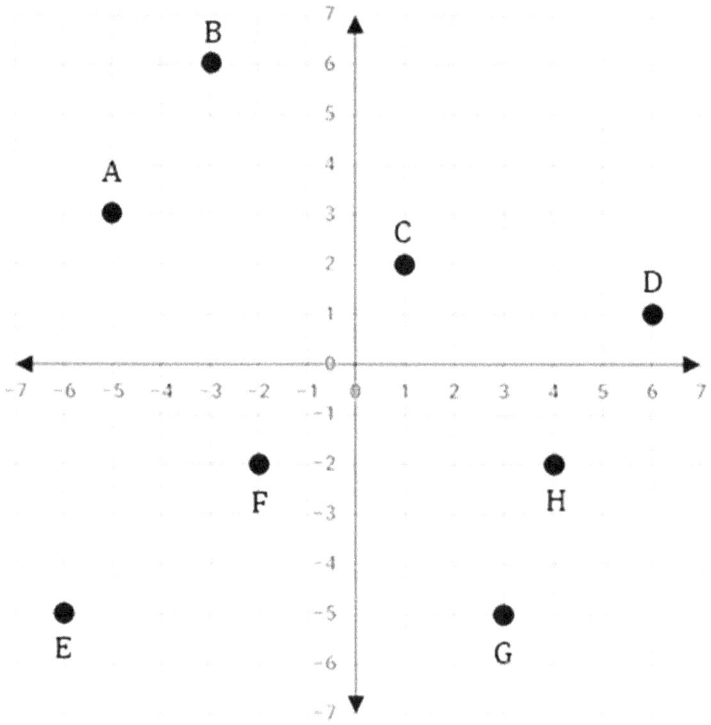

A. ________________ B. ________________

C. ________________ D. ________________

E. ________________ F. ________________

G. ________________ H. ________________

Workbook Answers

Chapter 1: Whole Numbers and Operations

Day 1

8-Digit Addition

1. 104,852,533
2. 150,350,925
3. 175,231,733
4. 152,742,958
5. 165,092,257
6. 164,329,602

Day 2

8-Digit Subtraction

1. 13,138,506
2. 30,680,207
3. 22,547,298
4. 35,657,655
5. 7,593,206
6. 46,473,155
7. 55,559,777
8. 89,277,747

Day 3

Multiplication Word Problems

1. 3,381
2. 956
3. 1,536
4. 608

Day 4

Multiplying 4-Digit Numbers by 2-Digit Numbers

1. 31,572
2. 68,740
3. 152,875
4. 32,147
5. 134,096
6. 174,801
7. 59,496
8. 226,023
9. 51,940
10. 128,808
11. 465,519
12. 382,932

Day 5

Multiplying 3-Digit Numbers by 3-Digit Numbers

1. 66,794
2. 69,689
3. 258,937
4. 128,853
5. 230,955
6. 130,104
7. 180,048
8. 103,170
9. 350,966

Day 6

Multiplying 4-Digit Numbers by 3-Digit Numbers

1. 2,133,216
2. 345,555
3. 1,143,539
4. 1,110,444
5. 2,901,440
6. 2,880,279
7. 1,093,680
8. 855,218
9. 1,770,131

Day 7

Dividing 5-Digit Dividends by 1-Digit Divisor

1. $4540\frac{1}{3}$
2. $15{,}337\frac{1}{2}$
3. $2{,}160\frac{3}{8}$
4. $8{,}502\frac{4}{7}$
5. $32{,}380\frac{1}{2}$
6. $10{,}381\frac{4}{5}$
7. $24{,}623\frac{2}{3}$
8. $10{,}136\frac{7}{8}$
9. 13,302
10. $8{,}324\frac{3}{4}$
11. $3{,}439\frac{5}{9}$
12. $14{,}152\frac{4}{7}$

Day 8

Division Word Problems

1. 52 Hours
2. 27 Miles
3. 51 Hours
4. 119 Sections

Day 9

Dividing 5-Digit Dividends by a 2-Digit Divisor

1. $2{,}280\frac{6}{29}$
2. $2{,}268\frac{5}{9}$
3. $1{,}278\frac{3}{43}$
4. $1{,}055\frac{13}{36}$
5. $1{,}357\frac{3}{8}$
6. $1{,}189\frac{27}{28}$
7. $459\frac{6}{35}$
8. $1{,}029\frac{5}{33}$
9. $1{,}345\frac{15}{44}$
10. $942\frac{87}{91}$
11. $1{,}050\frac{35}{62}$
12. $1{,}113\frac{9}{89}$

Day 10

Expanded Form to Exponent Form

1. 6^4
2. 3^3
3. 5^3
4. 7^4
5. 9^5
6. 8^6
7. 2^6
8. 4^4
9. 5^2
10. 6^5
11. 7^7
12. 2^3
13. 3^4
14. 9^6

Day 11

Exponent Form to Expanded Form to Standard Form

1. $9 \times 9 \times 9 = 729$
2. $5 \times 5 \times 5 \times 5 \times 5 = 3{,}125$
3. $8 \times 8 \times 8 \times 8 = 4{,}096$
4. $2 \times 2 \times 2 \times 2 \times 2 \times 2 \times 2 \times 2 \times 2 = 512$
5. $6 \times 6 \times 6 \times 6 = 1{,}296$
6. $4 \times 4 \times 4 \times 4 \times 4 = 1{,}024$
7. $8 \times 8 = 64$
8. $6 \times 6 \times 6 \times 6 \times 6 = 7{,}776$
9. $7 \times 7 \times 7 \times 7 = 2{,}401$
10. $3 \times 3 \times 3 \times 3 \times 3 \times 3 \times 3 \times 3 = 6{,}561$
11. $4 \times 4 \times 4 = 64$
12. $5 \times 5 \times 5 \times 5 \times 5 \times 5 \times 5 = 78{,}125$

Day 12

Comparing Exponents

1. $16 > 9$
2. $64 > 9$
3. $16 < 25$
4. $7{,}776 > 64$
5. $27 > 25$
6. $125 > 49$
7. $4{,}096 > 125$
8. $216 < 729$
9. $256 < 4{,}096$
10. $25 < 81$
11. $343 > 4$
12. $256 > 9$
13. $512 > 16$
14. $81 < 729$
15. $36 < 49$

Chapter 2: Integers and Operations

Day 13

Integers on the Number Line

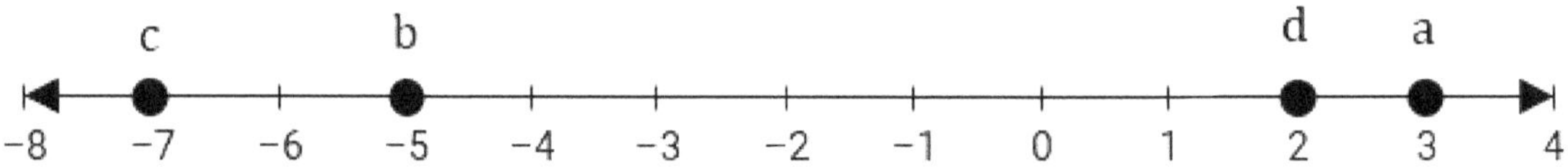

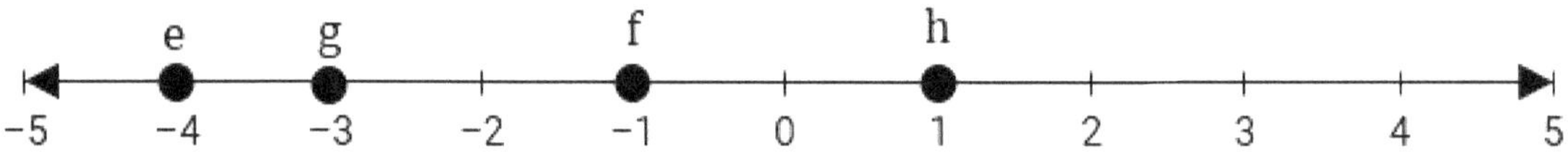

Day 14

Adding and Subtracting Integers

1.
   -3

2.
   -5

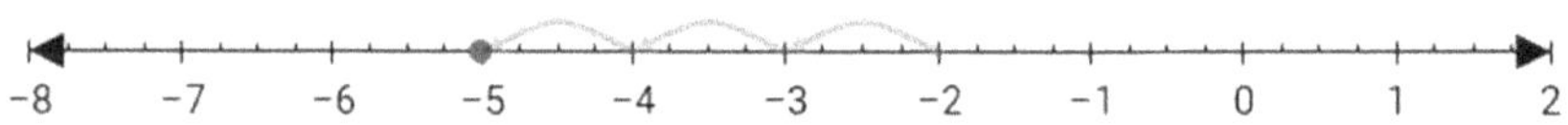

3.
   4

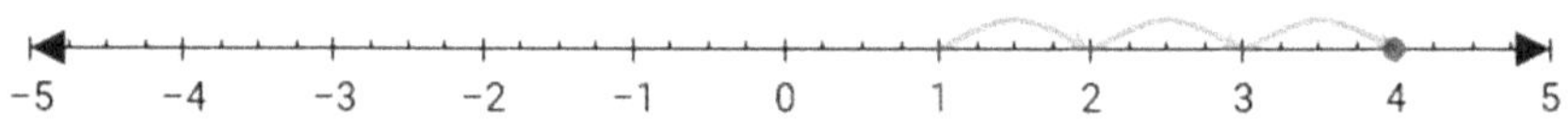

4.
   -7
5.
   -15
6.
   -1
7.
   -7
8.
   -7

Day 15

Multiplying and Dividing Integers

1. −21
2. −48
3. 16
4. −224
5. −900
6. 1,120
7. 7
8. −12
9. −4
10. 12
11. 234
12. −14

Day 16

Absolute Value and Opposites

1. 6
2. 8
3. 12
4. 13
5. 66
6. 49
7. −5
8. −23
9. 17
10. 62
11. −45
12. 98

Chapter 3: Rational Numbers and Operations

Day 17

Adding and Subtracting Fractions with Unlike Denominators

1. $\frac{25}{28}$
2. $\frac{3}{7}$
3. $\frac{3}{10}$
4. $\frac{1}{15}$
5. $\frac{5}{9}$
6. $\frac{2}{3}$
7. $\frac{31}{35}$
8. $\frac{41}{60}$
9. $\frac{31}{33}$

Day 18

Multiplying Fractions

1. $\frac{1}{7}$
2. $\frac{2}{5}$
3. $\frac{45}{64}$
4. $\frac{10}{21}$
5. $\frac{13}{42}$
6. $\frac{16}{45}$
7. $\frac{7}{48}$
8. $\frac{11}{42}$
9. $\frac{15}{26}$

Day 19

Dividing Fractions

1. $\frac{9}{10}$
2. $\frac{1}{9}$
3. $\frac{3}{4}$
4. $\frac{10}{21}$
5. $\frac{2}{3}$
6. $\frac{4}{7}$

Workbook Answers

Day 20

Converting Between Mixed Numbers and Improper Fractions

1. $\frac{7}{3}$ 2. $\frac{9}{2}$ 3. $\frac{44}{7}$

4. $\frac{35}{9}$ 5. $\frac{21}{10}$ 6. $\frac{23}{4}$

7. $2\frac{3}{4}$ 8. $1\frac{1}{6}$ 9. $4\frac{4}{5}$

10. $9\frac{1}{7}$ 11. $4\frac{1}{8}$ 12. $7\frac{1}{2}$

Day 21

Simplifying Fractions

1. $\frac{5}{11}$ 2. $\frac{4}{5}$ 3. $\frac{2}{3}$

4. $\frac{2}{3}$ 5. $\frac{7}{11}$ 6. $\frac{1}{2}$

7. $\frac{3}{4}$ 8. $\frac{4}{5}$ 9. $\frac{3}{4}$

Day 22

Converting Fractions to Decimals and Percents

1. 0.75, 75% 2. 0.25, 25% 3. 0.125, 12.5% 4. 0.4, 40%
5. 0.8, 80% 6. 0.2, 20% 7. 0.5, 50% 8. 0.6, 60%

Day 23

Converting Decimals to Fractions and Percents

1. $\frac{3}{10}$, 30% 2. $\frac{63}{100}$, 63% 3. $\frac{1}{2}$, 50% 4. $\frac{11}{20}$, 55%
5. $\frac{1}{4}$, 25% 6. $\frac{4}{5}$, 80% 7. $\frac{3}{8}$, 37.5% 8. $\frac{3}{4}$, 75%

Day 24

Converting Percents to Decimals and Fractions

1. 0.88, $\frac{22}{25}$ 2. 0.25, $\frac{1}{4}$ 3. 0.30, $\frac{3}{10}$ 4. 0.80, $\frac{4}{5}$
5. 0.75, $\frac{3}{4}$ 6. 0.50, $\frac{1}{2}$ 7. 0.35, $\frac{7}{20}$ 8. 0.99, $\frac{99}{100}$

Day 25

Arranging and Decimals Operations

1. 971.2
2. 522.16
3. 233.49
4. 480.18
5. 1,535.15
6. 338.301
7. 16.151
8. 6.905
9. 579.732

Day 26

Adding & Subtracting Decimals

1. 8,947.33
2. 14,437.96
3. 10,098.36
4. 9,065.53
5. 20,055.64
6. 14,216.45
7. 20.514
8. 109.208
9. 10,213.10
10. 1,588.44
11. 6,203.53
12. 5.35945

Day 27

Multiplying Decimals

1. 124.576
2. 14.5497
3. 1,656.86
4. 167.678
5. 42.526
6. 14.1327
7. 340.548
8. 110.451
9. 927.5

Day 28

Multiplying Decimals 2

1. 938.448
2. 357.7665
3. 449.709
4. 1.831536
5. 5696.46
6. 1,242.448
7. 318.8694
8. 304.3408
9. 1,371.806

Chapter 4: Expressions and Equations

Day 29

Determining the Variable

1. The total amount of money spent at the coffee shop each week.
2. The amount of money left to spend for the month's groceries.
3. The amount of ornaments left to make.
4. The amount of books that must be read each week.
5. The amount of pie each person will receive to eat.
6. The amount of money needed to save each month.

Day 30

Expressions Versus Equations

1. Expression
2. Expression
3. Equation
4. Equation
5. Expression
6. Expression
7. Expression
8. Equation
9. Equation
10. Expression

Day 31

Writing Expressions and Equations

1. $26 + 3x = 44$
2. $18 \div x + 9 = 15$
3. $31x \div 2$
4. $24 \div 3 = 8$
5. $72 \div 8 + 1$
6. $9 + (19 - 5x)$
7. $x(56 \div 7) = 9$
8. $10(66 - x)$

Day 32

Simplifying Expressions

1. 24	2. 18
3. 18	4. 10
5. $4 + x$	6. $7x$
7. $6y$	8. 8
9. $3x + 12$	10. $42 + 4z$

Day 33

Solving Equations

1. $x = 7$	2. $y = 16$
3. $x = 6$	4. $z = 24$
5. $x = 12$	6. $y = 7$
7. $a = 15$	8. $m = 15$
9. $x = 7$	10. $b = 6$

Chapter 5: Proportional Reasoning

Day 34

Ratios

1. 3 to 6, 3 : 6, $\frac{3}{6}$	2. 12 to 4, 12 : 4, $\frac{12}{4}$
3. 4 to 6, 4 : 6, $\frac{4}{6}$	4. 12 to 3, 12 : 3, $\frac{12}{3}$
5. 4 to 25, 4 : 25, $\frac{4}{25}$	6. 3 to 25, 3 : 25, $\frac{3}{25}$
7. 8 to 4, 8 : 4, $\frac{8}{4}$	8. 4 to 8, 4 : 8, $\frac{4}{8}$
9. 8 to 12, 8 : 12, $\frac{8}{12}$	10. 4 to 12, 4 : 12, $\frac{4}{12}$
11. 2 to 3, 2 : 3, $\frac{2}{3}$	12. 1 to 3, 1 : 3, $\frac{1}{3}$

Day 35

Rates and Unit Rates

1. Rate: $\frac{120 \text{ miles}}{4 \text{ hours}}$ Unit Rate: $\frac{30 \text{ miles}}{1 \text{ hour}}$
2. Rate: $\frac{\$12}{6 \text{ notebooks}}$ Unit Rate: $\frac{\$2}{1 \text{ notebook}}$
3. Rate: $\frac{30 \text{ pages}}{3 \text{ days}}$ Unit Rate: $\frac{10 \text{ pages}}{1 \text{ day}}$
4. Rate: $\frac{2 \text{ miles}}{20 \text{ minutes}}$ Unit Rate: $\frac{10 \text{ minutes}}{1 \text{ mile}}$
5. Rate: $\frac{\$8}{8 \text{ juice boxes}}$ Unit Rate: $\frac{\$1}{1 \text{ juice box}}$

Day 36

Proportions

1. $x = 4$
2. $x = 8$
3. $x = 10$
4. $x = 2$
5. $x = 2$

Day 37

Percent Proportions

1. $x = 10$
2. $x = 12$
3. $x = 10$
4. $x = 12$
5. $x = 18$

Chapter 6: Measurement and Geometry

Day 38

Finding Area of Triangles

1. $(11 \times 6) \div 2 = 33 \text{ in}^2$
2. $(12 \times 17) \div 2 = 102 \text{ ft}^2$
3. $(5 \times 14) \div 2 = 35 \text{ in}^2$
4. $(75 \times 46) \div 2 = 1{,}725 \text{ in}^2$
5. $(94 \times 89) \div 2 = 4{,}183 \text{ in}^2$
6. $(112 \times 156) \div 2 = 8{,}736 \text{ ft}^2$

Day 39

Finding Area of Parallelograms

1. $6 \times 6 = 36\ \text{ft}^2$
2. $13 \times 5 = 65\ \text{in}^2$
3. $15 \times 17 = 255\ \text{ft}^2$
4. $75 \times 45 = 3{,}375\ \text{ft}^2$
5. $136 \times 115 = 15{,}640\ \text{in}^2$
6. $563 \times 599 = 337{,}237\ \text{ft}^2$

Day 40

Finding Area of Trapezoids

1. $\frac{(6+9)\times 6}{2} = 45\ \text{yd}^2$
2. $\frac{(24+16)\times 8}{2} = 160\ \text{ft}^2$
3. $\frac{(20+32)\times 12}{2} = 312\ \text{in}^2$
4. $\frac{(55+110)\times 40}{2} = 3{,}300\ \text{ft}^2$
5. $\frac{(136+175)\times 212}{2} = 32{,}966\ \text{in}^2$
6. $\frac{(250+210)\times 88}{2} = 20{,}240\ \text{in}^2$

Day 41

Finding Perimeter

1. $12 + 4 + 4 + 6 + 8 + 10 = 44$ yd
2. $4 + 13 + 1 + 12 + 15 = 45$ ft
3. $88 + 200 + 96 + 88 + 105 + 88 + 411 = 1{,}076$ in
4. $8 + 8 + 8 + 8 + 8 + 8 + 8 + 8 = 64$ ft
5. $20 + 20 + 5 + 15 + 15 + 5 = 80$ in
6. $35 + 6 + 17 + 17 + 19 + 15 + 6 = 115$ yd
7. $60 + 55 + 34 + 27 + 10 = 186$ ft
8. $12 + 18 + 19 + 12 + 22 + 20 + 23 = 126$ in

Day 42

Volume of Rectangular Prisms

1. $3{,}600\ \text{ft}^3$
2. $300\ \text{in}^3$
3. $156\ \text{cm}^3$
4. $24\ \text{cm}^3$
5. $1{,}152\ \text{in}^3$
6. $48\ \text{ft}^3$
7. $3{,}600\ \text{in}^3$
8. $180\ \text{units}^3$

Workbook Answers

Chapter 7: Data Analysis

Day 43

Bar Graphs

1. 60
2. 35
3. 45
4. 30
5. 45
6. 50
7. 65
8. 25
9. 40
10. 20
11. 25
12. 5
13. 25
14. 5
15. 20
16. 30
17. 45
18. 10

Day 44

Histograms

1. The height of the black cherry trees in feet.
2. How many black cherry trees fall within each height range.
3. The 75-80 feet range.
4. About 10 trees.
5. Taller than 75 feet.
6. Most black cherry trees are between 70 and 85 feet tall.

Day 45

Pie Charts

1. Tennis shoes
2. Tennis shoes
3. Tennis shoes
4. Boots
5. Sandals
6. Tennis shoes
7. Boots
8. Boots
9. Boots
10. Sandals

Day 46

Stem & Leaf Plots

1. 19
2. 56
3. 94
4. 5
5. 77

Day 47

Box and Whisker Plots

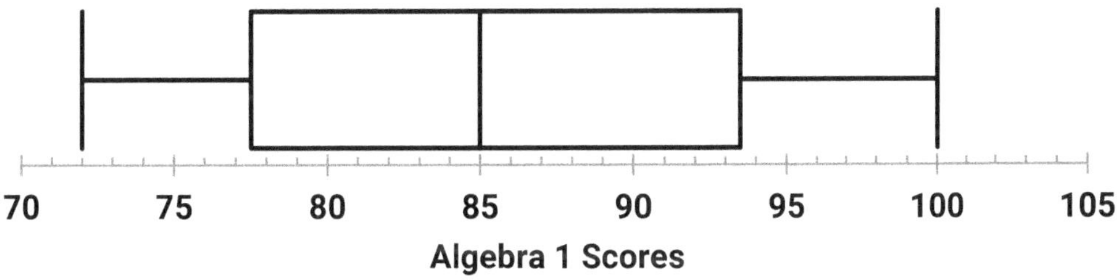

1. 72
2. 100
3. 85
4. 77.5
5. 93.5

Day 48

Mean, Median, Mode, and Range

1. Mean: 75 Median: 79 Mode: 79 Range: 75
2. Mean: 6 Median: 6 Mode: 5, 9 Range: 7
3. Mean: 26 Median: 26 Mode: 24, 26 Range: 10

Day 49

Graphing on a Coordinate Plane

1.

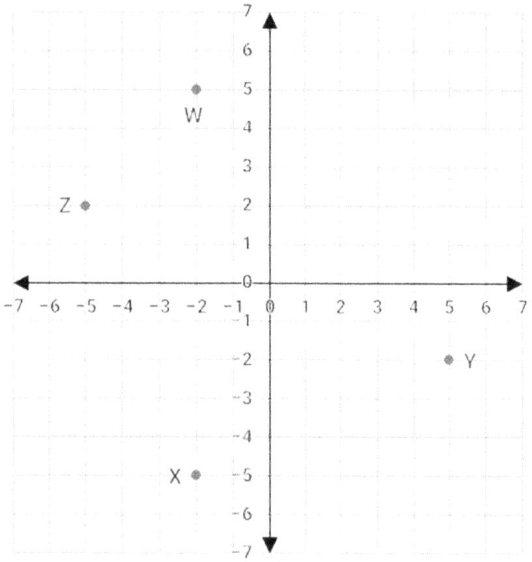

2.

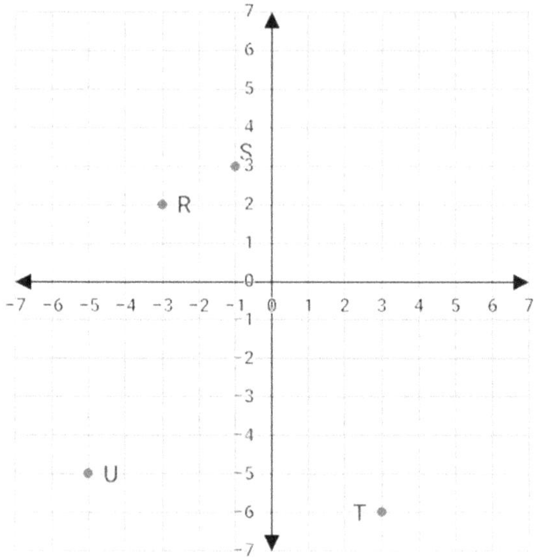

Day 50

Determining Ordered Pairs Using a Coordinate Plane

A. (−5,3) B. (−3,6)
C. (1,2) D. (6,1)
E. (−6, −5) F. (−2, −2)
G. (3, −5) H. (4, −2)

www.ingramcontent.com/pod-product-compliance
Lightning Source LLC
LaVergne TN
LVHW082357200926
848481LV00012B/3590